A NOTE ON THE AUTHOR

Laurie Winkless is an Irish physicist and author. After a physics degree and a masters in space science, she joined the UK's National Physical Laboratory as a research scientist, specialising in functional materials. Now based in New Zealand, Laurie has been communicating science to the public for 15 years.

Since leaving the lab, Laurie has worked with scientific institutes, engineering companies, universities, and astronauts, among others. Her writing has featured in outlets including Forbes, Wired, and Esquire, and she appeared in The Times magazine as a leading light in STEM. Laurie's first book was Science and the City, also published by Bloomsbury.

@laurie_winkless

Some other titles in the Bloomsbury Sigma series:

Sex on Earth by Jules Howard
Spirals in Time by Helen Scales
A is for Arsenic by Kathryn Harkup
Suspicious Minds by Rob Brotherton
Herding Hemingway's Cats by Kat Arney
The Tyrannosaur Chronicles by David Hone
Soccermatics by David Sumpter
Science and the City by Laurie Winkless
The Planet Factory by Elizabeth Tasker
Turned On by Kate Devlin
Borrowed Time by Sue Armstrong
The Vinyl Frontier by Jonathan Scott
Clearing the Air by Tim Smedley
Superheavy by Kit Chapman
The Contact Paradox by Keith Cooper
Life Changing by Helen Pilcher
First Light by Emma Chapman
Models of the Mind by Grace Lindsay
Handmade by Anna Ploszajski
Beasts Before Us by Elsa Panciroli
Our Biggest Experiment by Alice Bell
Aesop's Animals by Jo Wimpenny
Fire and Ice by Natalie Starkey
Racing Green by Kit Chapman
Wonderdog by Jules Howard
Growing Up Human by Brenna Hassett
Wilder by Millie Kerr
Superspy Science by Kathryn Harkup

STICKY

THE SECRET SCIENCE OF SURFACES

Laurie Winkless

B L O O M S B U R Y   S I G M A
LONDON • OXFORD • NEW YORK • NEW DELHI • SYDNEY

BLOOMSBURY SIGMA
Bloomsbury Publishing Plc
50 Bedford Square, London, WC1B 3DP, UK
29 Earlsfort Terrace, Dublin 2, Ireland

BLOOMSBURY, BLOOMSBURY SIGMA and the Bloomsbury Sigma logo
are trademarks of Bloomsbury Publishing Plc

First published in the United Kingdom in 2022
This edition published 2023

Figure 1: Illustration re-imagined by Marc Dando. Figure 2: © Youssef Grace,
Wikimedia Commons, CC BY-SA 4.0. Figure 3: © Gabriel Nodea and Robyn Sloggett.
Figure 4: © Laurie Winkless. Figure 5: © Wilhelm Barthlott, lotus-salvinia.de.
Figure 6: © Marc Dando. Figure 7: © Journal of the Royal Society Interface.
Figure 8: © Daly-Engel TS, et al (2018) doi: 10.3897/zookeys.798.28375,
Wikimedia Commons, CC BY-4.0, Christian Ferrer. Figure 9: © Wilhelm Barthlott,
lotus-salvinia.de. Figures 10–16: © Marc Dando. Figure 17: Mark Lincoln, NZ Raw.
Figure 18 © Rulsch, Wikimedia Commons, CC BY-SA 4.0. Figures 19–21:
Marc Dando. Figures 22 and 23 © Sile O'Modhrain and Alex Russomanno.
Figure 24: © Laurie Winkless. Figure 25: © NanoscaleRanger,
Wikimedia Commons, CC BY SA 4.0

A catalogue record for this book is available from the British Library

Library of Congress Cataloguing-in-Publication data has been applied for

ISBN: PB: 978-1-4729-5085-7; eBook: 978-1-4729-5081-9

2 4 6 8 10 9 7 5 3 1

Typeset by Deanta Global Publishing Services, Chennai, India
Printed and bound in Great Britain by CPI Group (UK) Ltd,
Croydon CR0 4YY

Bloomsbury Sigma, Book Seventy

To Richard. For holding my hand
through everything.

Contents

Hello 9

1 To Stick or Not to Stick 17

2 A Gecko's Grip 51

3 Gone Swimming 85

4 Flying High 117

5 Hit the Road 149

6 These Shaky Isles 179

7 Break the Ice 213

8 The Human Touch 247

9 Close Contact 279

Further Reading 313
Acknowledgements 327
Index 329

Hello

There's a flowchart lurking around corners of the internet. It is familiar to anyone who enjoys fixing and making things. At the top, it asks 'Does It Move?' and at the bottom, it offers two solutions: duct tape for when you want to hold something in place, and WD-40® for when you want to get things moving. These two products have long been seen as must-haves. Essential tools for any workshop; versatile enough to find frequent use. I'm a fan of both.

A few years ago, as the initial idea for this book was rattling around in my head, I realised something about these products. Because one sticks firmly on to surfaces, while the other slips between objects in order to loosen them, they're often viewed as opposites; as if they each occupy an end point of a stickiness-to-slipperiness scale. In reality, no such scale exists – not in our everyday lives, nor in the controlled environment of a research lab. That's because the words 'sticky' and 'slippery' are ambiguous, and certainly not precise enough to exist in opposition to one another. Though widely used, they mean different things to different people on different days. Depending on the situation, they might conjure up images of chewing gum, duct tape and sugary syrup on the one hand, or an icy road, WD-40 and wet tiles on the other. The words 'sticky' and 'slippery' are also not *true* materials properties in the way that, say, hardness and thermal conductivity are. They have no agreed-upon scientific definitions, and no specific metrics that can be used to quantify or compare them. That contrast – between the presence of these words in daily life,

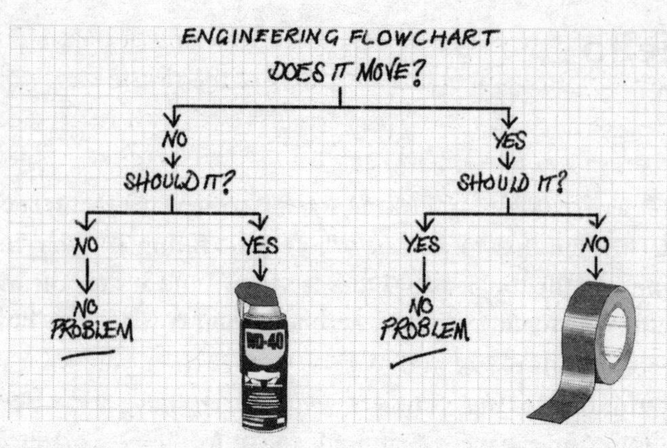

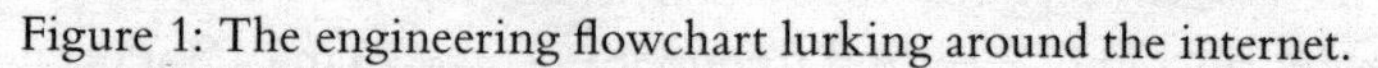
Figure 1: The engineering flowchart lurking around the internet.

and their absence from the scientific literature – is one of the reasons I decided to call this book *Sticky.*[*]

As I see it, this familiar term can be repurposed and applied to a vast array of interesting interactions: specifically, any of the weird and wonderful things that happen on and between surfaces. So much science happens where two things meet; whether that's air flowing over a curved surface, two pieces of metal sliding along one another, or glue applied to a plank of wood. And while stickiness isn't something that can be measured or defined, there are lots of other related properties that *are* measurable, and whole areas of research dedicated to defining them.

Tribology is one of these areas.[†] Sometimes described as the science of 'rubbing and scrubbing', its focus is on how moving surfaces interact with one another. While at first glance that might seem a bit niche, as we'll discover, such interactions are all around us, defining the movement of

---

[*] That, and I happen to think it's a pretty great title.

[†] The word tribology comes from Greek – 'tribos', which means 'I rub'.

glaciers on rocky landscapes and the whizzing of a hard-disk drive in your computer. Regardless of the sector they work in, something all tribologists are obsessed with is **friction**, the resistive force that acts parallel to surfaces, either to hold stationary objects in place (static friction) or to slow down the motion between those that move (kinetic friction).

By measuring the friction forces between materials, and incorporating them into mathematical models that have been developed and updated over decades, tribologists can glean a deep and sophisticated understanding of surfaces. In doing so, they can find ways to control the friction that acts on them. Every system with connected parts, be it engineered or biological, has been designed with friction in mind. Sometimes the aim is to maximise it; to provide grip or traction between components even in extreme conditions. Other times friction is the enemy, causing things to literally grind to a halt. Either way, we can't ignore it, which is why friction is at the heart of this book. It is the thread that runs through the fabric of every chapter.

In many ways, tribology is not a new science. Humanity has been exploring and manipulating surface interactions for millennia, far longer than we've had the equations or the tools needed to describe them. A famous example of this can be found in the burial tomb of Djehutihotep, a powerful provincial governor who lived in Upper Egypt 4,000 years ago. On the tomb's richly decorated walls is a mural now dubbed *Transport of the Colossus*. It depicts a huge monument atop a wooden sled, dragged by a team of hauliers. A lone figure standing at the foot of the monument can be seen pouring liquid directly in front of the sled, in what was initially interpreted as a purely ceremonial act. Engineers who later saw the image wondered if there was more to it. Could this liquid also be an example of an early

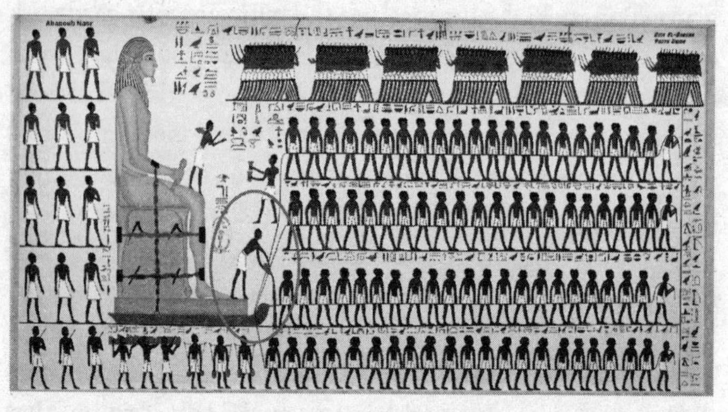

Figure 2: In this reproduction of the Djehutihotep mural, by artist Abanoub Nasr (working with the Deir Al-Barsha Youth Union), an individual can be seen pouring liquid directly in front of the sled.

lubricant; a way to make it easier to slide the heavy sled along the sand?

In 2014, a team led by Professor Daniel Bonn set out to answer that question. The experimental design was pretty simple – they'd load a small sled with weights, pull it along samples of sand that had been mixed with varying quantities of water, and measure the forces involved. The metric they were most interested in was the **coefficient of friction**, μ (pronounced 'mu'). This ratio is used a lot in tribology studies (and in engineering and science in general) because it gives you a clue as to how strongly two material surfaces interact with each other.* The closer its value is to zero, the more easily the surfaces can start to

* More specifically, μ is the ratio of the friction force that resists motion between surfaces, and the normal force (the 'supportive' or pressing force that a surface exerts on an object sitting on it). And just as there is both kinetic and static friction, there are different values of μ depending on whether the surfaces are at rest or in motion. We'll talk about μ a lot throughout the book.

slide. So steel-on-ice has a slightly lower μ than wood-on-ice (μ = 0.03 versus 0.05), whereas the frictional interaction of rubber-on-dry-asphalt is 18–30 times higher than either of them (μ = 0.9). This partly explains why tyres help vehicles to stay on the road; we'll cover much more on this in Chapter 5. By measuring the coefficient of friction of the sled being pulled along increasingly wet sand, Bonn could directly determine the effect that adding water had on the sand's 'slipperiness'.

Friction was high for all of the dry sand samples, with a typical μ of 0.55. Bonn attributed this to the 'heap of sand [that] forms in front of the sled before it can really start to move'. As he increased the water content, the size of that sand heap decreased, as did the value of μ. In some cases, friction between the sled and the sand dropped by 40 per cent, solely through the addition of water. But once the sand contained anything beyond about 5 per cent water, friction began to climb again, making the sled harder to pull. The researchers concluded that for transporting objects along desert sand, there is an optimal amount of water that can aid in sliding. The mechanism behind it will be familiar to anyone who has ever filled and flipped over a bucket to make a sandcastle. If the sand inside it is dry, it will flow and spread out freely. In contrast, wet sand can retain its shape, thanks to the formation of water bridges between the sand grains. If you get the mix just right, the water holds the material together, providing a smooth, stiff surface on which to slide heavy objects. Speaking to the *Washington Post* back in 2014, Bonn said that if this lubricating mechanism were scaled up to the projected size of the giant stone monument, it would mean 'that the Egyptians needed only half the men to pull over wet sand as compared to dry … the Egyptians were probably aware of this handy trick.'

The world of lubrication has largely moved on from using water. Today, there are thousands of lubricants available commercially, the majority of which are based on mineral oils (aka petroleum). What they all have in common is their aim: to reduce friction between moving surfaces, whether they're inside a cheap lawnmower or a high-tech Martian Rover. The global market for these friction-reducing compounds is enormous, worth in excess of US$150 billion (£107 billion) in 2020. We'll talk about some state-of-the art solid lubricants in Chapter 9. Water does still occasionally influence lubrication, especially in geological processes like landslides, and in the earthquakes and ice of Chapters 6 and 7. But more often than not, water, like many other fluids, exerts a friction force on surfaces. It drags on objects, slowing them down as they move through it. These particular resistive forces can be understood through **fluid dynamics** – the science of liquids and gases in motion – and their implications are widespread. As we'll discover in Chapter 4, the flight of every ball and every aircraft is controlled by the air around it. For the swimmers among you, Chapter 3 will uncover what it takes to slice through water, and you'll meet some underwater technologies that reduce water's influence by pushing it away from surfaces.

There are, however, lots of things that for various reasons didn't make it into this book. For example, something I'd originally planned to include was a chapter on the medical uses of surface science, from targeted drug delivery via engineered particles to designing implants that encourage cell adhesion and growth. Given that as I write these words (January 2021), the COVID-19 pandemic continues to impact the daily lives of everyone on the planet with a virus that can be transmitted by air and on surfaces, this omission is regrettable. But the truth is, I ran out of both time and space, for a topic that requires plenty of both. Other chapters have merely changed focus. Chapter 2 was going to explore the many ways that animals use surface science to navigate and control their surroundings. Spiders, sea-urchins and sharks were all on the list of possibilities. Instead, the chapter now focuses on just one animal – the gecko. In researching this lizard I became captivated by it: the astonishing mechanisms behind its climbing ability, and the many technologies it has inspired. There are other examples from the natural world scattered throughout the rest of the book. In Chapter 8, I've taken a physicist's perspective on our sense of touch, and of its role in human society. And finally, or perhaps, 'firstly', Chapter 1 is an introduction to all things adhesion, including descriptions of how some of the sticky and slippery products that I'm frequently asked about actually work.

At its heart, *Sticky* is a book about materials, and the forces at play on their surfaces. In one way or another, I've been professionally interested in this topic since 2007. That's when I first got involved in a research project into the use of patterned surfaces to control both friction and fluid flow, which led to work on water-repellent materials, among other things. Later, when I was writing *Science and the City*, these surface interactions just kept popping up, from the slipperiness of leaves on the railway line to the grip of tyres to the road. The importance of friction to the modern world seemed laughably outsized compared to our knowledge or appreciation of it. That's really when the idea for *Sticky* first took hold. Once I started seeing things from the point of view of stuff-that-happens-on-surfaces, I couldn't stop. This book is the result.

Sticky is not intended to be an exhaustive exploration of all known surface interactions. Nor is it trying to be a physics textbook, a mathematical treatise on friction, or a deep-dive into the best glues on the market. If that's the level of knowledge you're looking for, there are lots of other references that I will happily point you to.* Instead, what you'll find within these pages are my favourite examples of how the forces that act on the outer skin of materials can literally and figuratively shape the world around us. The implications of these forces cut across scientific disciplines, and as a result, our journey will take some surprising twists and turns. I think (hope?) that there's something in here for everyone.

In researching these topics, I've been privileged to speak with an array of fascinating people from across science and society; all experts in their respective fields, they generously gave their time to talk to me and share their knowledge. To say 'I owe them' would be an understatement. I'm excited for you to meet each of them.

So why not slip into something comfortable, stick on the kettle, and let me tell you some stories.

---

* There's a Further Reading section at the end of the book with a selection of key references. And you'll find the full reference list (with links, where possible) on my personal website.

CHAPTER ONE

To Stick or Not to Stick

The far north-west of Australia might not seem like an obvious place for a book on surfaces to begin. But if we want to explore humanity's connection to all things sticky, there's nowhere better. Famed for a dramatic landscape of steep gorges and pristine waters, the Kimberley region is vast – five times the size of Ireland – yet it is home to fewer than 37,000 people.* It is also indescribably ancient, formed at least 1.8 billion years ago, and largely left alone by tectonic forces since. Its soil varies from bright yellow through countless shades of red and, occasionally, to a purple so deep it looks black.

The region's sunset palette is a result of different forms of iron hydroxide in its rocks, with each combination of iron, oxygen and hydrogen atoms producing its own hue. Collectively, these materials are referred to as ochre, humanity's first pigment. For millennia, the Kimberley's Aboriginal inhabitants have masterfully used ochre to make their mark: to share stories, honour their ancestors and reflect their experience of the world around them. Today's artists may paint on canvas or wood, but their work forms an unbroken line back to the earliest form of artistry – rock art. And the Kimberley is home to some of the finest, and oldest, examples on the planet.

---

* This is equivalent to 0.08 persons per square km. The population density of Australia as a whole is 3.2 persons per square km.

Arguably the most famous are the Gwion-style motifs. Found in the northern Kimberley, they're described as being 'dominated by finely-painted human figures in elaborate ceremonial dress including long headdresses, and accompanied by material culture including boomerangs and spears'. Despite their immense cultural value, many Gwion sites have been damaged or destroyed, largely through development. Gaps in the cultural heritage protection legislation are widely blamed, with Simon Hawkins from the Yamatji Marlpa Aboriginal Corporation describing current protections as 'archaic … a joke'. It's understandable, then, that the people of the Kimberley have been cautious about sharing their ancestral knowledge. In the book *Gwion Gwion: dulwan mamaa*, written by four senior elders (munnumburra) of the Ngarinyin people, Gwion rock art was described as 'a secret to protect man … blood … law'.

In more recent years, however, many Aboriginal communities have sought help from western scientists to understand how and when this art was created. Australian rock art is notoriously difficult to date, because its iron-based pigments lack carbon, the essential element used in radiocarbon dating. But a study published in 2020 found an inventive way around that. Working alongside Traditional Owner groups, University of Melbourne scientists studied Gwion artwork from 14 sites at (deliberately) undisclosed locations. From each site, they took tiny samples of wasp nests that were either under or over the paint used to create the images. By carbon-dating these nest remnants, the researchers could provide a lower and upper limit on the age of each artwork. They concluded that most of the Gwion motifs they studied 'were painted over a relatively narrow time span between 11,500 and 12,700 years ago'.

Though ancient, this piece of rock art is far from the oldest example in Australia. That title is currently held by a sample found on Jawoyn country in the Northern Territory in 2012. A small piece of quartzite bearing painted charcoal shapes, understood to be a fragment of a much larger painting, was dated at 28,000 years old. And there's plenty of other archaeological evidence to suggest that sites like these have been occupied for much, much longer.* But that topic is a book in itself. My own fascination surrounds the remarkable staying power of ochre. How could a paint – applied to a rock wall 23,000 years *before* the building of Egypt's oldest pyramid – have stuck around for so long? And what's its relationship to the high-tech, molecularly controlled paints of today?

I was lucky enough to be granted an interview with Gabriel Nodea, esteemed artist and senior knowledge-holder of the Gija people. Gabriel's paint-making process fuses traditional and modern materials. Like his forefathers, he grinds brightly coloured rocks to create his powdered pigments. But for a binder – the liquid that holds the pigment together and helps it adhere to surfaces – he uses PVA glue mixed with water. His paint is robust and can last for many decades on a canvas, but he says 'it wouldn't stick on a rock wall. I can't really tell you how they did it. We only have clues so it's very, very hard to describe. All I know is that they were using their eyesight and their minds, and looking at things from a different point of view. They

* An archaeological study on tools found at Minjiwarra – a large sedimentary feature in north-eastern Kimberley – concluded that this region has been continuously occupied by Aboriginal communities for 50,000 years.

Figure 3: Gija artist Gabriel Nodea with one of his pieces that was painted using ochre. This piece describes the story of Warmun.

must have had some secret ingredient because just water and ochre doesn't work.'

'Researchers have tried for a long time to identify the binding agents used in rock art,' said Dr Marcelle Scott, a Research Fellow at the Grimwade Centre for Cultural Materials Conservation and a colleague of Nodea's. 'The main challenge is the small amount of material you have access to. The other thing – and this is likely to have contributed to its resilience – is the chemical similarity between the art and the rock surface itself.' Scott, speaking to me over the phone from Melbourne, said the latter can lead to some interesting conclusions: 'People are very quick to say "blood" when they see iron oxide in a sample. Most of the time, it has come from the ochre.' Blood *is* sometimes used in Aboriginal art – Jack Britten, a famed Gija artist who passed away in 2001, was known to bind his ochres with gum-tree sap and a dash of kangaroo blood. But, so far as I could find, it hasn't yet been definitively identified on any of Australia's traditional rock art, and that's true for most other potential materials. Sites in other parts of the world have given up some of their secrets; for example, traces of sap from aloe vera plants have been found in the paints used in South Africa's San rock art. But most of the chemical information we have on these ancient paints comes from analysis of their pigments. One study looked specifically at a distinctive mulberry pigment used in Gwion rock art. Using portable X-ray fluorescence to analyse small samples of the ochre, researchers showed that its vibrant colour was due to jarosite – a mineral that contains potassium and sulphates. Other studies have done everything from identifying cleverly disguised historical vandalism, to pinpointing the exact quarry a specific ochre was mined from.

The fact that ochre pigment was, and still is, 'of the land' was made especially clear in 2011, when a small community in East Kimberley was struck by tragedy. For more than 50 years, Warmun Art Centre has held a special place in the world of contemporary indigenous art. Owned and managed by the Gija people, it has produced some of Australia's most celebrated artists, and it acts as a vital repository for cultural knowledge and artefacts. So when flash floods hit the region, devastating the Art Centre and the homes around it, its impact was profound. 'There were paintings everywhere; they were spread so far we had to go looking for them on motorbikes,' Gija artist Roseleen Park told the *Sydney Morning Herald* at the time. 'They were up trees, on hills, wrapped around barbed wire. I rescued around a hundred or so.' Nodea, who was then Chairperson of the Centre, said that his overriding feeling was of loss: 'It was devastating to see our paintings washed away and damaged. All the art throughout Australia is very, very important for our people – it keeps us connected with our culture and with strong stories, connected to country, connected together.'

Some of those damaged works came under the care of Scott and her colleagues at the University of Melbourne. 'The Warmun Community Collection, which features the work of deceased artists, is nationally significant, so we were deeply honoured to get involved as conservators.' But, she said, the challenge they faced was enormous. 'Hundreds of these pieces were caked in mud, and many were mouldy.' The conservation team also had to contend with a range of different substrates – everything from wood and canvas to cement sheet. And, as we'll discover in the pages to come, how well a paint bonds to a surface depends just as much on what it's sticking to as what it's sticking with. In the end, Scott said, the paint surfaces proved to be incredibly robust. 'Dealing with difficult surfaces is part of our job, but I was worried about our ability to remove the mud – which in reality is wet ochre – without damaging the painting underneath. But the Warmun artists really know their materials. With their help, we were able to clean them much better than I anticipated.' The carefully restored artworks were returned to Warmun in 2013, and are now housed in a purpose-built, elevated storage facility close to the Art Centre.

Unlike these contemporary works, the art that has adorned the walls and caves of the Kimberley for millennia can't be isolated from the destructive forces of nature. And those forces can be extreme. The region's wet season is stiflingly hot and humid, while the dry season is characterised by bright sun and night frosts. The dramatic climate makes the survival of this ancient rock art all the more remarkable. But as it turns out, one type of weather might actually have helped to *preserve* them.

Desert varnish is a dark, thin (> 0.2mm) coating that can form on a range of exposed rock surfaces.* Though most common in dry, arid regions, it's been found everywhere from Iceland to Hawaii, and it tends to be rich in manganese and iron oxides, similar compounds to those found within the rocks on which it forms. What differentiates the coating from the rocks underneath is high concentrations of silica and aluminium, as well as a host of other oxides. These minerals transform it into a hard, glassy surface that protects the rock, and according to scientists, the only way they could have got there is via the wind. As it rolls its way across the desert, wind picks up particles of dust, and often deposits them on rock faces. What happens once it gets there remains a bit of a mystery. Some sort of biological mechanism involving tiny spores of fungi has been suggested, as has the chemical breakdown of silica in the presence of water. One thing we do know is that the rate at which desert varnish forms has varied through time and across sites. One Australian study found evidence of a period of 'major varnishing' at least 10,000 years ago, followed by multiple distinct layers of different thicknesses.

According to Scott, processes like these could explain why some rock art has survived for so long. 'Weather definitely contributes, but its impact depends on when an event happened, relative to the age of the painting. A particularly wet epoch soon after the piece was completed would destroy it, but a dry epoch might lend itself to the formation of a protection layer.' Even when it does form,

* This is equivalent to 200μm, which is about the same thickness as two sheets of copy paper.

desert varnish is not impervious. Salt spray and fire have both been shown to destroy it, and in some regions, the rate of destruction is outpacing formation. It's unclear what impact, if any, this might have on the long-term health of Australia's precious rock art sites. Bundled together with other factors, like climate change, mining operations and population growth, it paints a worrying picture.

I pondered all this as I stared at an overhanging rock covered in layers of hand prints and stencils – a sea of orange, white and the red that gives this place its name. The Red Hands Cave, an hour west of Sydney, is considered one of the best examples of Aboriginal art in the Blue Mountains. These images, formed by spraying ochre over the hands and palms of young boys as part of their initiation into adulthood, may have been on this wall for 1,600 years. I couldn't see any sign of desert varnish, but the pigments were still bright. I edged forward, pressing my nose against the metal fence that protects the site from vandalism, eager to get as close as possible. For several minutes, the only sound was the rustle of the gum trees. As I headed back towards the track, I passed some other visitors and heard a small voice say, 'Dad, we could put handprints on the wall beside my bed. It'd be so cool!' The response was swift: 'Maybe, love, but don't expect them to last as long as these ones.'

Paint

Modern paint still largely follows the model of a solid pigment (which provides the colour) suspended in a liquid medium (which adheres the pigment to the surface). But it also includes a range of different additives, each conferring specific properties onto the final product. The result is a seemingly infinite array of paints, tailored for every need. From bridges and cars to glass and canvas, if there's a surface that needs a coating, there's a product that will stick to it. To understand how paints achieve this, let's first talk about how they're made.

If you're in the business of producing artist's oil paint, you'd start with a medium, and the most common one is linseed oil. This clear, straw-coloured oil is produced from the seeds of the flax plant.* Although linseed oil might seem sufficiently 'sticky' to cling to a canvas, a stabiliser – typically a compound called aluminium stearate – is also added. Next, it's time to slowly incorporate a finely powdered pigment of your choice. For a bright white, you'll need titanium dioxide; for a nice blue, a dash of cobalt aluminate. Using a large mixing blade, churn this mixture for up to four hours. This allows the stabiliser to find its way to the pigment particles, coating them and helping them to disperse through the oil. It also works to thicken the mixture, giving it a more paste-like texture. The proto-paint's next destination is the triple-roll mill: three huge rollers made from granite or stainless steel, spinning at different speeds, separated by just a few micrometres. The mill pulls the paint through increasingly small gaps, to grind and separate the pigments ever more finely. In his book *Chromatopia*, master paint-maker David Coles writes that the mill 'is the heart of paint-making', but warns that using it is as much a craft as a science. Friction between the rollers heats them up, causing them to expand, changing the size of the gap. This increase in temperature can change the paint's flow behaviour, as can slight differences between batches of pigment. As a result, 'The paint-maker must be constantly attentive to the

---

* The edible form of linseed oil is flaxseed oil. It's processed without the use of solvents.

vagaries of milling.' The results, though, are spectacular: a smooth, vibrant, buttery paint that – depending on the colour – can be more than 50 per cent pigment, by volume. Once it passes tests for texture and its colour is checked against that of previous batches, the paint is packaged up into tubes, ready for use.

Something that oil paints don't do on surfaces is *dry*, because drying involves losing moisture to the air. Instead, oils actively remove oxygen from the air and use it to form bonds between adjacent molecules. As the reaction – known as curing or polymerisation – continues, it creates a dense network of identical, interconnected molecular chains that form solid films, which work to gradually harden the paint. This also means that oil paintings actually get heavier as they cure.* In fact, linseed oil can increase its weight by more than 15 per cent, depending on how the oil was produced.

Waterborne paints – more typical of those used by DIY-ers to 'freshen up' a room – are very different. As the name suggests, their pigments are suspended in water. Once applied to a surface, this liquid gradually evaporates, leaving behind a thin film of pigment held together by binder compounds. The ratio of pigment to medium is generally much lower in waterborne paints than in oils. Colin Gooch, Technical Director of paint manufacturer Resene, told me that 'a 4-litre can of a high-quality waterborne paint might contain just over 1.5 litres of volume solids – that's what actually forms the film that stays on the surface. The job of the other 2.5 litres is to keep those solids dispersed. It allows us to carry the pigment from the can to the wall.'

---

* … at least initially. Eventually other processes take place that work to release some of the compounds from the oil into the atmosphere.

The process of making a standard can of paint is much more industrial than artisan, but as I saw for myself, that doesn't make it any less impressive. Resene is a Kiwi icon – a 100 per cent-New Zealand-owned business that's been making paint for more than 70 years. At its factory and Head Office in Naenae, 20km outside Wellington, Resene manufactures all of its waterborne products. And having used a small fraction of them to redecorate our home, I was very keen to peek behind the curtain, and learn a bit of paint chemistry along the way. It became immediately clear that with Gooch, I was in good hands. Despite having been the company's senior chemist for half a century, an enthusiasm for paint seems to bubble out of him. Shaking my hand after meeting in the lobby, he said, 'I really should warn you. The paint industry is addictive. Once you're in and you see the challenges and the opportunities, you get hooked!'

After a coffee and a brief chat about the company's history, we walked down to the labs, and started talking about the challenges of making paint. Gesturing towards a huge array of labelled samples on shelves, he said, 'First off, there's no such thing as a perfect paint. Nothing we can make will stick to everything, because a concrete surface is very different from the weatherboards of a house. If we want to produce a successful paint for a specific substrate, we need to have an intimate understanding of that substrate. Once we have that, we can start thinking about the formulation.'

Pigments are usually step one. Just like those used in oil and ochre paint, these pigments are powders, and their size and shape offers lots of opportunities to an aspiring chemist. Titanium dioxide, which Gooch referred to as a 'very coarse pigment', has an average particle size of just 300nm (0.0003mm) – you could fit close to 3,500 of them into a poppy seed. And individual particles of their (much finer) magenta pigment have 26 identifiable sides.* These nanoscale complexities are important because generally, the more available atoms there are on a surface the easier it is to get them to react. 'And for paint, we need to trigger a chemical reaction,' said Gooch, as we crossed the yard and headed towards a large building. 'Take sugar dissolved in water. If we let the water evaporate, we'll be left with the same sugar grains we started with – it hasn't undergone any change.' The analogy is a good one because, like sugar, most commercial pigments are hydrophilic, a term that literally translates as 'water-loving'. Getting them to dissolve in water is not an issue. What is an issue, is keeping them stuck on a surface as that water disappears. He continued, 'We want a pigmented film that will actually adhere onto a substrate, and that means playing with the particle's surface chemistry.'

For this, paint-makers look to binders, often called resins, made from a wide array of compounds. Gooch is a fan of acrylic polymers, largely because of the long, heavy molecules they can form. 'The higher the molecular weight, the more durable a paint will be,' he said. 'The reason is simple. Most aggressors to paint durability – one example is UV light – work to break down molecules. The bigger the starter molecule, the longer it'll take to degrade.' Acrylics can also be blended in different quantities to change the hardness of the final paint film, and the process is quick and efficient. 'Acrylics technology is particularly elegant, and it's hard to beat,' Gooch enthused.

---

* A polygon of this shape is called an icosihexagon. Just a fun little titbit for your next quiz.

They do have one problem, though. Acrylics are hydrophobic; they repel water. 'So how do you get them to stay in the water medium?' I asked, confused. In my mind, I pictured tiny acrylic particles floating towards the top of a bowl, as they're repelled by the H_2O molecules in all directions.

'The interface between the hydrophobic binders and the hydrophilic pigments is the greatest weakness in paints,' he replied. 'We strengthen it through an intermediary – surfactants or amphiphilic* molecules, which have two distinct ends. One's attracted to the binder and the other to the pigment.' These molecules act as bridges, chemically connecting two particles that would otherwise never meet, and they create droplets that can happily float around in water. When you apply this mixture to a surface, and the water begins to evaporate, these pigment-binder droplets get closer to each other and to the surface, eventually forming a tough, elastic film. Otherwise known as a painted patch on your wall.

'A huge amount of paint engineering goes solely into the transitory stage, just to keep it stable in the can. The surfactant can be a real pain on the surface, but we'd struggle to get our paint onto the surface without it!' As Gooch said this, we were both looking into an enormous vat of a bright white liquid. A let-down tank, I was told, holding 10,000 litres of standard white paint. As well as the water and pigment-binder particles, the tank likely held other paint additives – perhaps something to confer mould resistance, or to make the final film easier to clean. From here, samples of the almost-completed paint would be

---

* You are extremely likely to have a bottle of amphiphilic molecules in your home – they're better known as detergents. They help to lift hydrophobic oils and fats from surfaces, a job that water can't do alone.

tested in the lab and checked against references. After getting the all-clear, the paint would then be canned and moved to the on-site warehouse for distribution. At this site the entire process, from raw ingredients to a sealed can of paint, takes around two days.

Adhere

Among all this talk of paint, I've omitted something important – the mechanism by which any adhesive material (e.g. a coating, paint or glue) actually sticks to a surface. In other words, what is adhesion, really? And how can we best define just how 'stuck' something is?

The answer to those questions really depends on who you ask. To a chemist, adhesion might be best described in terms of energy; to a physicist or engineer, adhesion is all about forces. Both views are valid because, on a fundamental level, adhesion is the attraction that exists between molecules of dissimilar materials. And the strength of that adhesion is usually defined by the effort or work it takes to separate them: either the amount of energy or the maximum force required to dismantle the connection.

Entire careers have been dedicated to the study of adhesion, and the global industry for adhesives is enormous, worth US$45 billion (£33 billion) in 2018. So, attempting to summarise it in a few pages may well be a fool's errand. Regardless, let's start with the basics – understanding what a connection between materials might look like. Consider a generic, if artificial interface: a drop of an unnamed liquid sitting on a clean, solid block of material. The general consensus is that there are three or four ways that these materials could interact:

Chemically. This happens when molecular bonds form between an adhesive and the surface. In paint, this sort of adhesion is enabled by the binder molecules that surround the pigment particles. They react with molecules on the surface, sharing and borrowing electrons, effectively forming a new compound at the interface. This is associated with a process called adsorption, whereby the adhesive needs to 'wet' the surface – more on this shortly.

Mechanically. No solid material is ever truly smooth. Even a highly polished sheet of glass is, at the microscale, a jumble of peaks and valleys. The theory goes that if a liquid can flow into all of these irregularities, it can make an especially intimate contact with the surface. No reactions actually occur between the liquid and the solid. This connection – called interlocking – is physical rather than chemical. The liquid grips onto the surface in a similar way to a rock climber sticking their fingers into cracks and crevasses. This mechanism is why paint manufacturers suggest using sandpaper to roughen your wall or wooden furniture before painting it: the process adds more peaks and valleys to your surface. Roughness is also widely believed to act as a barrier to cracks that might otherwise propagate between a surface and its coating.

However, the importance of this mechanism may well depend on what the product is there to do. Paint is a coating, so it's a liquid applied to a solid surface that will dry or cure in place. In contrast, adhesives are used to join things together – they're the 'meat' in a sandwich, rather than the top layer of bread. And if you're trying to bond two materials with a liquid adhesive, roughness might not always lead to mechanical interlocking between materials. Professor Kevin Kendall, who's a big name in adhesion science, wrote a book called *The Sticky Universe*.* In it, he describes roughness as 'always acting against you, whether making or breaking the joint', and shows that in some cases, increasing surface roughness can significantly reduce the adhesion between two materials.

But whether you're gluing or painting, there is one outcome to roughening a surface that everyone seems to agree is beneficial. In the act of sanding or abrading a material you indirectly clean it; you change its surface chemistry by removing oils, greases and other contaminants that build up over time. A clean surface will always offer better adhesion than a contaminated one, and that's true regardless of how rough your interface is. So it's not that roughness *doesn't matter* to adhesion – it just isn't the only thing that matters.

To understand **diffusion**, a third way in which a liquid-solid interface can interact, I called on Professor Steven Abbott, an adhesives guru and Fellow of the Royal Society of Chemistry. As someone who has worked in industry for decades, he's well-versed in the differing demands placed on adhesives and coatings.† And over the course of a long, early morning call, he explained that the function of a product will define the relative importance of each adhesion model.

Diffusion generally only happens when the solid in question is a polymer, which doesn't mean it's rare. Polymers are everywhere, both in the natural world (rubber, silk, cellulose) and the manufactured world

* Kendall is one of three people who developed the Johnson-Kendall-Roberts (JKR) theory. It describes how two bodies deform when in contact with one another.

† In 2020, Steven Abbott published a book with the Royal Society of Chemistry. Called *Sticking Together*, it talks about the science of adhesives and paints in much, much more detail than I've managed here.

(nylon, silicone and Teflon™). What makes something a polymer is structure: its molecules are arranged into long, repeating chains. In this model of adhesion, molecules don't bind to each other as much as intermingle across the interface, like two plates of cooked spaghetti combined into a single pot. In the paints and coatings industry, diffusion is seen as a fairly niche process, one that doesn't really apply to their products. But, Abbott said, it plays a vital role in adhesives.

> People dismiss the importance of diffusion, because they fixate on the idea that bulk polymers can't mix. But they forget that fifty years ago, a scientist called Eugene Helfand showed that polymers follow very different thermodynamic rules at an interface. There, you can easily get a few nanometres of stuff intermingling. And often that's enough to give you a strong joint between materials.

According to 3M™, global giant of sticky products, there is also a fourth model of adhesion: **electrostatic** interactions. The company says that if you've ever seen a piece of paper move towards some sticky tape before you were ready to place it, you've experienced this effect yourself. But I'd argue that this attraction – enabled by charged particles that build up on the tape as you pull it from the roll – isn't actually what *adheres* the tape onto the surface. Yes, it helps draw the materials together, but it's not what holds them there. So I'm a bit hesitant to call electrostatics a 'model of adhesion'. Perhaps the confusion arises from the fact that there is a related attractive force that can operate between closely packed atoms: van der Waals forces, which are often grouped in with electrostatic interactions, despite being subtly different from them. These tiny interatomic forces certainly play a role in adhesion, but as we'll discover in Chapter 2, it's not humans who have managed to exploit these particular forces to their fullest.

None of these models, on their own, can fully explain adhesion, and for any given adhesive product, it's almost impossible to determine exactly which model(s) might be operating. As Monique Parsler, then Chief Chemist at Bostik New Zealand, told me, 'Every class of adhesive works on a different principle, or combination of principles.'* There are other factors at play too, like **cohesion**, which is a liquid's ability to stick to itself. You can think of it as an inner strength that comes from bonds that form between like molecules. For a paint or adhesive to be durable, it needs to have both good *cohesion* and good *adhesion*. If either one of those fails, the product will fail too. In paint, a cohesive failure might look like film that has been stripped of its colour, whereas an adhesive failure involves the paint physically lifting off the surface. Either way, your surface will need to be repainted.

Energy

There's something else we need to discuss – a very large elephant in the surface science room. That is the concept of **surface energy**. This is a measure of the excess energy on the surface of solid materials that results from those outer atoms having unbalanced bonds compared to atoms deep within the bulk. It's a real, measurable property, and its value gives you an idea of how attractive a material surface is to other molecules. One way to visualise surface energy, which

* For interactions between two solid materials, the list of adhesion models is even longer, and includes things like magnetism (which is outside the scope of this book).

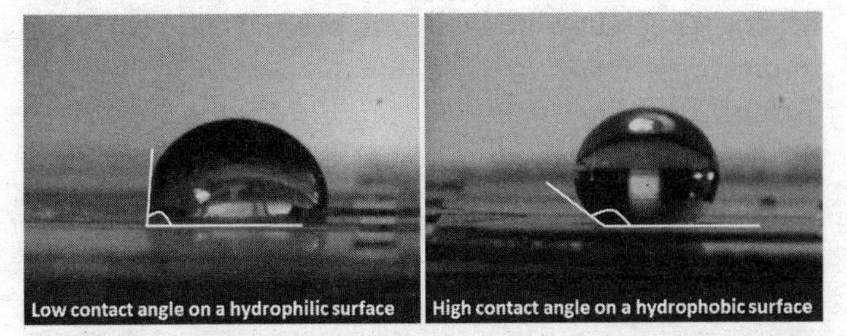

Figure 4: The contact angle of a water droplet on a surface tells you something about that surface. If the contact angle is low (left image) the surface is hydrophilic, or water-attracting. If the contact angle is high (right image) the surface is hydrophobic or water-repelling.

is sometimes called wettability, is to watch how liquids interact with that surface. You probably instinctively know that a droplet of water will behave very differently on a block of wood, a non-stick frying pan, a waxy leaf and a piece of cardboard. On some of these, the droplet will immediately spread out, whereas on others it'll bead up into a ball. By measuring the angle that the edge of a droplet makes with a surface (known as the contact angle, θ, or theta), you can calculate a value of surface energy. And if the liquid you've used is water, this measurement can also tell you how hydrophobic (water-repellent) or hydrophilic (water-loving) the surface is.

These definitions are on a sliding scale, and the boundaries between them are blurry, but generally, surfaces that have a water contact angle of between 0° and 90° are said to be hydrophilic. These materials are strongly attractive to liquid molecules, which means their surface energy is high – they wet easily. If you measure a contact angle of between 90° and 180°, you have a low surface energy material. When a water droplet is deposited on these surfaces, it is largely 'ignored'. That droplet tends to hold its shape rather than spread out, which means that the surface is considered to be hydrophobic. We'll come back to these very soon.

With paint, you want your liquid to spread, because in theory, that will tap into several adhesion models at once. A surface that wets easily is a surface that can be painted without too much stress, so knowing the wettability of a surface is extremely useful. As an added bonus, these measurements can also tell you something about the cleanliness of your surface, as the presence of contamination will change the contact angle, making it higher or lower, depending on the kind of contamination – another consideration when good adhesion is the goal. Surface energy also tends to correlate with the coefficient of friction, μ (we met this earlier, in the *Introduction*). It's not a hard-and-fast rule by any means, but if a material has low surface energy (if it is slippery to liquids), it is often also low-friction (slippery to solids).

Where conversations about surface energy get heated is around its importance in adhesive joints: two objects bonded together by a liquid. If you look at the website of every major adhesive manufacturer on the planet, it won't take you long to find a mention of surface energy. The property, typically measured in Dynes per centimetre, is very often used to assign materials into groups. Metals tend to have very high surface energy (for example, copper is 1,103 Dynes/cm) so they wet easily, and the values for traditional materials like wood tend to be quite high, in the tens to hundreds of Dynes/cm range. Engineered plastics like PVC and nylon have lower values, between 30 and 50 Dynes/cm. On most sites, the accompanying text explains that the further you go down the scale, the harder a material is to bond with.

On my visit to her lab at Bostik, Monique Parsler showed me a huge matrix that helps users choose an appropriate adhesive for a host of surfaces. For her and her colleagues, wetting is critical. 'When I'm looking at a new product, or at an old product for a new application, I want to get the lowest possible contact angle so that the surface is entirely wet. If you don't get surface wetting you will not get adhesion. It's as simple as that.'

Professor Abbott's views on it are rather different. He believes that while the principles of adhesion were worked out a long time ago, the general understanding of it – even by companies who make these compounds – is somewhat lacking. 'Surface energy is undoubtedly useful for paint, but for practical adhesive systems, where we're actually sticking stuff together, it is basically irrelevant, thousands of times too small to give us what we need. And yet people are obsessed with it.'

On his YouTube page, Abbott has a collection of videos largely targeted at industry. Nestled among them is a short demo on the limitations of surface energy-based adhesion. In it, he and another man can be seen trying to pull apart – tug-of-war style – two super-smooth, super-clean sheets of rubber. There's no adhesive between them; they're bonded purely by surface energy. Despite their best efforts, the rubber sheets do not budge, which is impressive. And yet, a young girl (Anna) easily peels them apart. 'Surface energy forces are great when it's a pure vertical pull, but in the real world you can't rely on having that,' explains Abbott. 'When Anna pulls the sheets apart, she's effectively forming a crack at the interface. Surface energy offers almost no resistance to that sort of force, so the adhesion is instantly lost.'

For Abbott, the ability to absorb and dissipate **crack energy** is the most important – and most overlooked – contributor to reliable adhesion. 'People think of strength and elasticity as being in opposition to one another, so they add more and more crosslinks to their adhesive to drive up its strength. But in almost all cases, this makes the adhesive worse, because it can no longer move and stretch.' In an overwhelming majority of cases, a level of flexibility endows an adhesive with resilience, a means of coping with a wide variety of stresses and strains. Or as Kendall once put it, 'Soft materials stick best.' Without that ability to absorb energy, an adhesive is very likely to fail, no matter how traditionally 'strong' it might be.

Towards the end of my conversation with Abbott, I asked him why he thinks surface energy is still held up on a pedestal in the world of adhesives. He sighed and said,

> Sometimes, I think it's just because you can measure it. It's a real number; and that usually makes people feel as if they're in control, and that they understand what's going on. There's a commercial drive too. A colleague who works in surface treatments once told me that his customers are only interested in the Dynes he's selling them, even though he tries to explain that there's much more to sticking than surface energy.

Given all of these factors, how can we know if we have a good adhesive or not? Well, 'good' is a relative term. Water might be able to stick a coaster onto the bottom of a glass, but you wouldn't want to rely on it to hold a plate together. And if sticky tape instantly created a permanent bond, wrapping birthday gifts would become a high-stakes activity. When choosing an adhesive, we're mainly looking for a material that's able to resist the forces that will act on it during its lifetime. Exactly what those forces look like will vary hugely depending on our specific needs. And vitally, there's no way to separate out an adhesive's behaviour from that of the surfaces it's sticking to, because as Abbott likes to say, 'Adhesion is a property of the system.' What that means in real terms is that everything depends on everything else. This is why there is no simple, objective measure of stickiness; no single number that will encapsulate all that you need to know about a product. The best we can do with commercial adhesives is to design tests that reflect how those products will be used in the real world.

There are a lot of adhesive products on the market, as well as a huge number of measurement standards and recognised test set-ups that put them through their paces. So rather than try (and fail) to go through everything, I've picked out the two products that I think everyone is familiar with.

I really couldn't write a book called *Sticky* without mentioning Post-It® Notes, especially because as I type, I am surrounded by them, each one covered in scrawled ideas for upcoming chapters.* Originally patented by 3M and first released onto the market in 1980, these sheets of coloured paper backed with a small strip of adhesive are an office mainstay. Given that Post-Its are now so commonplace, it's easy to forget how much engineering went into their development. The story is the stuff of design legend, and it involves two scientists: Spencer Silver and

---

* An aside: most companies are keen to talk about the technical intricacies of their work. Some require a little bit of coaxing, and/or the signing of a non-disclosure agreement. But despite my best efforts (more than 18 months of regularly contacting press offices, sending messages on every social media channel I could find, and even calling in favours from friends with connections to the company) I could not get an 'in' to 3M.

Arthur Fry. Back in the late 1960s, Silver had been working on ultra-strong adhesives for the aerospace industry, but a mistake in the lab led him to a discovery that he'd go on to patent – a sprayable, mildly sticky adhesive comprised of tiny acrylate spheres (each one between 5nm and 150nm, or 0.005mm to 0.15mm) suspended in a solvent. These spheres were pressure-sensitive but resilient, or as Silver wrote in the patent, 'A force applied directly to one of the polymer spheres will deform it; however, the spherical shape is reassumed upon release of the stress.' Going on to discuss how this material could be applied to different surfaces, he described it as 'a tacky adhesive layer which readily bonded paper, but permitted the paper to be removed, repositioned and rebonded'. It was several years before this barely sticky adhesive would find a commercial use. Fry, a colleague of Silver's and a keen member of his church choir, was frustrated that his carefully positioned bookmarks would frequently fall out of his hymn book. Looking for something that could stick to the page but be easily removed, he thought of Silver's invention. The two began collaborating and gradually assembled a team.

An issue they had to solve early on was that each time a proto-sticky note was removed from a surface, it would leave behind some of its polymer spheres, and be less sticky as a result. If they wanted the notes to be truly reusable, they'd need to find a way to keep the adhesive on the note paper. Their solution was effectively a glue for their glue – a binder compound, applied to the notes before the adhesive spheres, which anchored them to the paper. In the patent for their 'acrylate microsphere-surfaced sheet material', the researchers stopped short of naming the specific binder they used, as well as the mechanism by which it works (there's a mention of a 'vacuum effect', but that's it). They described the spheres as being 'partially embedded in and protruding from' the binder. The result is a pebbled, pressure-sensitive adhesive film that stays firmly stuck onto the note, but which holds the note onto a surface with a very small amount of adhesive force. From there, they had to design the machinery capable of mass-manufacturing sticky notes. 'It's not just a matter of smearing a little glue on the paper,' Fry was quoted as saying in *The Chemical Engineer.* The early prototype of the kit, understood to have used rollers to apply the binder followed by the adhesive spheres, was built in Fry's basement. Even after solving all the engineering challenges, the team still had to convince 3M management that the notes could be commercially viable – and that's a story in itself. Once launched worldwide, under the trademark Post-It, Fry and Silver's invention became immensely popular, and inspired other manufacturers to create their own versions. In 2019, the global market for sticky notes was estimated to be worth more than US$2 billion (almost £1.5 billion).

The other sticky superstar I want to highlight is superglue, which, it may surprise you to know, is not a trademark.* Most adhesive brands sell a product using this name, and all of them are based on polymers called cyanoacrylates. They're famed for their ability to stick to seemingly any surface, though that wasn't initially seen as a positive by the man who would go on to patent their use as adhesives. During

* As far as I could find, Henkel were the most recent company to hold the trademark 'Super Glue', but it was abandoned in 2010. There is also a separate trademark for 'The Original Super Glue', which is held by Pacer Technology®/Super Glue Corporation.

the Second World War, Eastman Kodak chemist Harry Coover was tasked with producing transparent gunsights for use by the military. While trialling different polymers, his team came across an exceptionally sticky formulation that stuck to – and permanently ruined – everything it touched. Interesting, yes, but because it wasn't what they needed, Coover set it aside. It wasn't until six years later, while researching adhesives for use in jet aircraft, that he revisited cyanoacrylates. In 1956 he was granted the patent for them.*

In the bottle or tube, cyanoacrylate adhesives are liquid, and they flow and behave as such. But as anyone who has ever accidentally stuck their fingers together will tell you, once they're out of their container, they very quickly turn solid. Contrary to popular belief, it's not the oxygen in the air that kick-starts this curing reaction; it's the water vapour. As soon as superglues are exposed to water, H_2O molecules bond to the cyanoacrylates, joining together to form long, interconnected chains that set hard. Most surfaces on our gloriously damp planet are permanently clad in an ultra-thin layer of water, making cyanoacrylate a very useful, very versatile option for your adhesive needs. It does also mean that skin, which generates its own water layer through respiration, is especially vulnerable to its speedily forming bonds. This realisation later led to cyanoacrylate compounds being used to close wounds

---

* Patent US2768109, granted to Coover, H.W. 1956. Alcohol-catalyzed α-cyanoacrylate adhesive compositions. The product has gone by lots of names since Coover patented it. Originally called 'Eastman #910', it was quickly rebranded as 'Superglue'. When Loctite later bought the technology from Kodak, they called it 'Loctite Quick Set 404', followed by 'Super Bonder'.

instead of traditional stitches, usually under trademarks like Dermabond, or SurgiSeal. As someone who had no sense of danger or self-preservation as a kid, I can attest to their effectiveness.

Adhesion is undoubtedly complex, but humanity's understanding of it has led us to some profound, as well as technically sophisticated discoveries, from painted masterpieces that last millennia, to a glue for every possible need. But something that is perhaps less obvious is that the same mechanisms that stick things to surfaces can also be used to *stop* them from sticking, and unsurprisingly, nature got there first.

Slip

I stared at the green, dinner-plate-sized leaves floating in the pool of a climate-controlled room. The light caught on water droplets held like jewels in the centre of some of the leaves, but otherwise they were pristine. My colleague Andrés and I were visiting the labs at the Royal Botanic Gardens at Kew in leafy west London, to learn more about the species of semi-aquatic plant that was now in front of us: *Nelumbo nucifera*. Better known as Indian Lotus, or simply lotus, it was a plant that even someone as devoid of botany knowledge as me had heard of. Sacred to both Hindus and Buddhists, the lotus is usually associated with purity, thanks to its ability to emerge unsullied from muddy, murky waters.

The secret to the lotus's eternal cleanliness lies on its surface, which was first scientifically described in the early 1990s by a German botanist, Professor Wilhelm Barthlott. For years, Barthlott and colleagues had been using an imaging technique called scanning electron microscopy (SEM) to study cacti, orchids and other subtropical plants. With a resolution significantly higher than that of a standard optical microscope, SEM uncovered a host of previously unknown structures – bumps, hairs and folds – on plant leaves. Botanists began to wonder if there might be a connection between those structures and the water-repellent behaviour that they'd observed in some species. By combining SEM imaging with contact angle analysis, and looking at the leaves of 340 different plants, Barthlott could begin to answer that question. He found that the majority of wettable leaves (those with low contact angles) were microscopically smooth, but tended to be dirty even after being rinsed with water. In contrast, the hydrophobic leaves were covered in wax crystals, making them microscopically rough. They were also usually free of contamination.

Most impressive of all were those leaves that combined a wax coating with a variety of microstructures. They had such high contact angles that they were considered superhydrophobic, and the lotus leaf was found to be the most superhydrophobic of all ($\theta = 162°$). Its distinctive hierarchical structure – densely packed, rounded features of differing sizes, all coated in a layer of rough, robust wax crystals – offers a significant barrier to any material that might want to stick to it. Water droplets can't penetrate this dense forest of microstructures. The very best they can do is rest on the top, in the form of a near-spherical droplet, making almost no real contact with the leaf. A subtle shake or tiny tilt is enough to send it rolling, and any dust that might also be present on the leaf is soon picked up by the water and carried away. Barthlott dubbed this non-stick, self-cleaning ability 'the lotus effect', a term he would later go on to trademark.

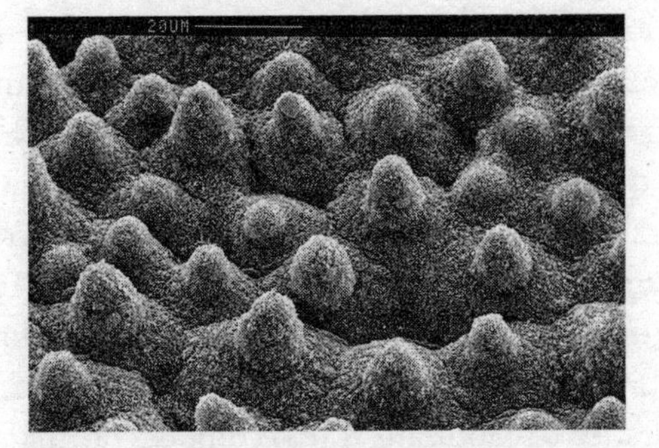

Figure 5: This image of a lotus leaf shows the complex patterns of tiny bumps that covers its surface. The scale bar at the top of the image is 20 μm or 0.02 mm long.

Since then, more than 9,500 papers have been published on the lotus effect. My own contribution to the field was very minor, and came right at the start of my scientific career. At the time, I had been tasked with investigating the wetting properties of engineered surfaces – silicon wafers that had been carefully etched to produce a range of different micro- and nanoscale patterns, and then coated with a polymer. In effect, we were taking inspiration from the lotus leaf, trying to see if we could use surface texture to make an already hydrophobic surface even more hydrophobic. So that day at Kew, where we got to see those pristine leaves up close, was eye-opening. It was a reminder that whatever incremental improvement we might look to achieve, nature had already mastered it. In the end, while some of our results on engineered surfaces were a bit inconclusive, we found that we could indeed control a material's water-repellent properties, simply by altering the size and shape of its surface features. In the most extreme examples, we saw contact angles of 86° and 154° on two chemically identical samples. All that differed between them was the minute pattern etched on their surfaces.

The lotus effect, and the interdependence between surface texture, water repellence and reduced contamination, have gone on to inspire the development of technologies like self-cleaning glass, stain-resistant fabric and anti-fungal paint. A recent project that caught my eye was TresClean, which wrapped up in 2020. Funded by the EU, TresClean focused on producing ultra-slippery anti-microbial metal and plastic surfaces for use in the food industry and in home appliances. 'We're looking at real components that biofilms are known to form on,' Dr Adrian Lutey told me over the phone from the University of Parma, Italy. These biofilms – slimy layers of microbes such as bacteria and fungi that can build up on surfaces in wet conditions – are extremely common both in the natural world and in industry. They're the cause of smelly breath and dental plaque. They can also slow down your washing machine or clog up pipes in water-treatment plants. Biofilms usually start with a single microbe attaching to a surface, so in theory, if you can prevent that from happening, you can stop these films from forming. The TresClean team looked at two species of bacteria that can pose risks to human health: *Escherichia coli* (*E. coli*) and *Staphylococcus aureus* (*S. aureus*), which have very different geometries and surface chemistries. *E. coli* cells are rod-shaped, up to 3nm long, and are surrounded by a thin fluid membrane. *S. aureus* cells are spherical, with a diameter of less than 1nm, and they have no outer membrane.

They then looked at how those bacteria, suspended in a fluid, would interact with a series of different surfaces: some untreated, some polished mirror-smooth, and others covered in textures produced by firing a laser at them. 'The laser is highly specialised,' explained Lutey. 'It produces ultrashort pulses that last less than one-trillionth of a second, and it can provoke some very interesting modifications on metal surfaces.' Microscale features including spikes, pillars and parallel ridges could all be generated by the laser. It's this final, ridge-like texture – officially called laser-induced periodic surface structures (LIPSS) – that proved to be the most effective at stopping bacteria from sticking. When comparing untreated stainless steel surfaces to those patterned with LIPSS, Lutey and his colleagues saw a drop of 99.8 per cent in *E.coli*, and 84.7 per cent in *S. aureus*. 'We're fairly certain that the LIPSS performed well for *E. coli* because of the dimensions of the surface structures. They're much smaller than the bacteria size, so the available contact area is reduced. It's like the bacteria cell is sitting on a bed of nails.' Perhaps surprisingly, the wettability of the surface made almost no difference to *E.coli* – the same low number of bacteria stuck around on hydrophobic surfaces as hydrophilic. But, Lutey said, their results with *S. aureus* were less clear. 'We don't have a convincing explanation for why we saw those reductions, but given that *S. aureus* didn't like our superhydrophobic spikes either, we expect that both surface wettability and topography play a part.' Even with a few unanswered questions, the results seem promising. Of course, it's hard to predict where it'll go from here. Lutey's hope is that TresClean's industrial partners – which includes BSH, the largest manufacturer of home appliances in Europe – will add the technology into their production lines. Who knows, maybe a dishwasher that keeps itself clean could be on the cards.

But it's likely that your kitchen already contains a very interesting slippery surface, perhaps the most famous of them all: polytetrafluoroethylene (PTFE), trademarked and better known as Teflon™. Like superglue, Teflon was an accidental discovery, first found on the inside of a gas canister by chemists experimenting with new refrigerants. Contrary to legend, this waxy, white solid isn't a by-product of the Apollo space programme. Rather, its corrosion-resistant properties saw it used in the Manhattan Project of the 1940s. Yes, Teflon aided in the development of the first atomic bomb. It was another 10 years before the material would be applied to cookware, though its specific formulation has changed since then.* Teflon's slipperiness comes purely from its polymer chemistry, rather than any lotus-like nanobumps or engineered ridges. Its long molecular chains consist of a carbon (C) backbone surrounded by fluorine (F) atoms, with the bonds between them described as 'the strongest … in organic chemistry'. This also leads to a high level of cohesion between Teflon molecules. In practical terms, this makes Teflon wholly unattractive to other molecules; or, as Steven Abbott put it, its 'fluorocarbon groups dislike everything in the universe that isn't fluorocarbon'. Compounds applied to a Teflon surface have no opportunity to react with it; they can't penetrate its structure or intermingle with its polymer chains. In effect, they ignore the surface, which is what makes it so non-stick. Unsurprisingly, its surface energy is also exceptionally low – 18 Dynes/cm according to 3M – and the coefficient of friction, μ, of Teflon-on-Teflon is just 0.04.

You might be wondering, then, how this super-non-stick material can be made to stick onto other surfaces, such as the aluminium that makes up the bulk of your frying pan.

* In the 1990s, the US Environmental Protection Agency ordered studies into the potential health risks of two of Teflon's original ingredients (called PFOS and PFOA). Since 2014, these ingredients have been considered 'emerging contaminants', effectively banning their use.

There have been several methods patented over the years, but from what I could find, they fall into two main categories. One is based on mechanical adhesion, and it starts by sandblasting the aluminium or bathing it in acid to roughen the surface. A thin primer coat of Teflon is then sprayed onto it, which rather than reacting with the surface, gets trapped inside the tiny holes and cracks created by the first step. Once baked at high temperatures, the Teflon solidifies in place. Additional layers of Teflon are added and baked, and because they can form chemical bonds with one another, they create a robust coating. The second way to make Teflon stick to other materials is to chemically treat the Teflon itself. You can do that either by bombarding it with charged particles to knock out some of its fluorine atoms, or use a compound that breaks some of the C–F bonds and replaces the fluorine with something else. Either way, you're left with exposed carbon atoms that are desperate to bond to something. Press the treated Teflon into an aluminium surface or any number of primer materials, and those carbons will happily and firmly stick. Baking the now-coated metal seals the deal. On most cookware, the Teflon coating measures 20–40nm; that's thinner than a sheet of copy paper. There's a lot of surface science in a humble frying pan.

Teflon has found a use in everything from dentistry and rain-gear to solar panels and air filters, all thanks to its ability to stop stuff from sticking. In recent years, other even lower-friction materials have appeared on the market. BAM, a composite made from boron, aluminium and magnesium ($AlMgB_{14}$) and titanium diboride (TiB_2) has a μ less than half that of Teflon's, as do some diamond-like carbon films. But, in terms of their cost and versatility, none look set to steal Teflon's crown of slipperiness any time soon.

Our knowledge of surfaces has been honed since time immemorial, from utilising the clays of the Earth to make our mark, to recognising the perfection of a self-cleaning leaf. We've taken these lessons and used them to craft materials that can control friction and manipulate the complex interactions between surfaces and fluids. We can create, build, join, enhance and beautify objects through clever design and chemistry. To my mind, there's no doubt that surface science shapes our world.

CHAPTER TWO

A Gecko's Grip

I met my first gecko in 2014, on the balcony of a Cambodian hotel. After a busy day of exploring in sweltering heat, I'd picked up some food and a cold beer from a stall, and headed back to my room to consume both, while overlooking the busy streets of Siem Reap. I quickly realised I wasn't alone – a 25cm-long lizard, pale grey with orange spots, clung motionless on the rough brick wall behind me. Several frantic Google searches later, I'd identified it as a Tokay Gecko (*Gekko gecko*), which is harmless to humans. So I sat back and enjoyed the company. Over the course of an hour, my balcony-mate moved up and down walls at remarkable speed, crossed the tiled floor, and at one point, scampered along the face of the glass patio door to munch on a sizeable spider. By the time I went to bed, it had taken up residence on the painted ceiling.

I'd known that geckos were famously good climbers, but what amazed me that night was just how adaptable this lizard was. Smooth or rough, painted or 'natural', no surface seemed too challenging for it, while we humans struggle to walk on icy streets (discover why in Chapter 7) and can't scale steep slopes without specialised equipment. The gecko's ability to cling to almost any surface has fascinated philosophers and scientists for millennia, and studies into that ability have regularly featured in scientific journals since the 1800s. A huge part of the lizard's mystique – and its genius – comes from the fact that its feet are not what you would think of as sticky. They're dry to the touch, and unlike the adhesives of the previous chapter they don't leave behind any gloopy residue. Geckos stick without stickiness. And it wasn't until recent decades that scientists finally understood how they do it. Lots of ideas were cast aside on that long road to discovery, but despite being disproven, a few seem to have stuck around (ha!). So let's do some debunking.

Ideas

If you look at the underside of a gecko foot, one of the first things you'll notice is that its toes are covered in flat, overlapping, scale-like ridges. These are called **lamellae**, and for at least a century, they were considered the main means by which geckos stick. In a book published in 1830, zoologist Johann Wagler proposed that lamellae act as suckers. The idea received widespread support at the time, and you can understand why. Several species of sea creature were already known to use suckers to cling to rocks and surfaces, and humans had been using straws to enjoy their drinks since at least 3000 BC. There was a general appreciation of the power of suction, which would eventually lead to several patents on practical rubber suction cups as we'd know them today.*

Like the gecko's feet, suction-based devices can grip without using a sticky substance. Under ideal conditions, they can also support considerable weight. Just ask 'Skyscraper Man', Dan Goodwin, a climber who since 1981 has used suckers to scale tall buildings. When you press a suction cup onto a surface, a little pocket of air is

* One of which, patented by an inventor called Orwell H. Needham in 1868, was delightfully called the 'Atmospheric Knob'. Search for 'US82629' on any web browser to see the knob in all its glory.

expelled out of the sides before the pliant, rubbery material forms a seal. This creates an area of low air pressure – a partial vacuum – inside the cup, while outside, you have normal atmospheric air pressure. The weight of the outside air molecules exerts a force on the cup's surface, but because there are far fewer air molecules inside it, they push back with a much lower force. The overall result is that the cup is firmly held onto the surface, where it'll remain for as long as the seal stays closed. Could lamellae operate in the same way?

A century after the idea was first published, a scientist named Wolf-Dietrich Dellit set out to test it. His hypothesis was that, if a gecko's feet truly adhered to surfaces via suction, they should behave like standard suction cups, which are less effective at lower air pressures. Demonstrating how little consideration was given to animal welfare in Nazi-era Germany, Dellit put live Tokay Geckos into a vacuum chamber, and slowly pumped the air out. Unlike a suction cup, the gecko's feet stayed stuck to the wall of the chamber, even in the vanishingly low air pressures of a vacuum, and beyond the point of its death. It was a convincing (if tragic) experimental outcome. The suction hypothesis was briefly revisited in 2000, when a group led by Professor Kellar Autumn – who has studied gecko adhesion for decades – managed to quantify the strength of the adhesive force between a gecko foot and a smooth surface. It was shown to be many times higher than could be achieved by suction, permanently shutting the door on that particular idea.

Another popular theory emerged in the first half of the twentieth century, enabled by improvements to the design of optical microscopes. Researchers realised that rather than being smooth, the lamellae on a gecko's toes were actually covered in tiny, densely packed hairs, which they called **setae**. And because these setae all appeared to be slightly curved and orientated at the same angle, people wondered if they might work like tiny hooks, allowing the gecko to grip onto irregularities on surfaces. That 'climber's boot' hypothesis – that setae acted as microscopic versions of the crampons used by mountain climbers – is now known as micro-interlocking, and again, it proved to be rather popular. Related to it was the suggestion that static friction may be involved. All of those hairs hugely increased the contact area between the lamellae and a surface, so perhaps that would also drive up the frictional force, helping the gecko to hold on.

It turned out to be relatively easy to test both ideas. If setae really were micro-hooks, you'd expect geckos to cling to rough surfaces more strongly than to smooth surfaces. Experiments from several research groups showed that geckos can not only scale surfaces so smooth that the largest 'bump' is just a few atoms in size, but in most cases, they adhere *more strongly* to smooth than to rough surfaces. So that's micro-hooks out. If static friction were to blame, a gecko attempting to cross a ceiling would fall off almost instantly. But observations in the wild suggest that geckos spend a lot of their time inverted, and according to Kellar Autumn, who spoke to me over the phone from Portland, 'The grip of a Tokay Gecko is so strong that, with all four of its feet in contact with the ceiling, it could support more than 30kg.' In the lab too, geckos have been observed walking on traditionally 'low-friction' surfaces like silicon. So that's friction out (for now).

But if it's not suction, friction or micro-interlocking that gives geckos their amazing climbing ability, what does that leave us with?

Charge

The aforementioned Dellit had another idea – that geckos might stick using electrostatic attraction. When you put two dissimilar materials in contact with each other, a curious thing happens. The surface of each material becomes electrically charged, one positively and one negatively, because of the wholesale movement of electrons from one surface to the other. As a result, these materials become attracted to each other. This is exactly the same mechanism that keeps a balloon stuck to a wall after you've rubbed it vigorously in your hair, and is what causes horrible crackling static between a wool jumper and a polyester shirt. Dellit reasoned that if he could eliminate this build-up of charges, he'd be able to test whether the gecko really adhered by electrostatics. So, adding to his notoriety, he used X-rays to strip electrical charges from the air molecules in a sealed chamber, neutralising electrostatic effects, and no doubt also exceeding the safe X-ray dose for the live gecko contained within. Despite this bombardment, the gecko clung on, leading Dellit to conclude that electrostatics were not what made the gecko a super-climber.

But this is one hypothesis that never entirely went away. In 2014, researchers at the University of Waterloo carried out a series of experiments to examine what role, if any, these charges played in gecko adhesion. The team took five Tokay Geckos and placed their feet onto ultra-smooth vertical surfaces coated in one of two polymers. As the foot pad of a gecko was brought into contact with each material, charges were seen to build up – positive on the foot pad, and negative on the polymers. They also measured the force required to drag that foot across each material. The higher the density of surface charges, the more tightly stuck the gecko foot seemed to be, which led the authors to conclude that 'electrostatic interactions ... dictate the strength of gecko adhesion'.

'This experiment was interesting, but I just don't buy their argument,' said Autumn, when I asked him about the implications of the paper. 'Once you start using whole animals, rather than individual setae, it's difficult to isolate effects from each other, and that makes it tricky to interpret exactly what's going on.'* While Autumn disagreed with the team's conclusion that electrostatic forces dominate gecko grip, he did concede that they might be an additional force for geckos to call on when faced with particularly slippery surfaces. On this point, other gecko experts, including Villanova University Assistant Professor Alyssa Stark, agree. 'It seems likely that there are lots of things going on at once,' she said. 'While most teams agree on the dominant force that geckos utilise, from our research I can't say that it's the *only* force at work. It's perfectly possible that electrostatics also play a role.'

Part of the complexity around gecko grip results from the fact that not all geckos are equal, or at least, they're not all equally sticky. Though we think of them as tropical animals, the thousand or so members of the family *Gekkonidae* have proven themselves to be incredibly adaptable, and can occupy a huge variety of habitats. One of

---

* Most of Autumn's experiments are carried out on small groups of hairs that have been removed from gecko feet. I've been assured that this can be done painlessly ... or at least that it's no more painful than tweezing out an errant eyebrow hair. Gecko setae also grow back within a couple of days.

the most famous, the Black-eyed Gecko (*Mokopirirakau kahutarae*), lives in the alpine mountains of New Zealand's South Island, while the Western Banded Gecko (*Coleonyx variegatus*) can be found in some of the most parched deserts in the US. As a result, each species of gecko is unique, having been forced to adapt to its surroundings in order to survive. Stark told me that this diversity makes it difficult to write hard-and-fast rules for geckos. 'Many species have claws, but some don't. Several gecko species have only three functioning toes, compared to the usual five. And then there's the enormous range of toe size and shape – the list of differences can be rather long.'

But among all those differences, there seems to be one mechanism that most geckos, well, adhere to. Yes, we're finally getting on to what it is that makes geckos stick.

Toes

The first clue to this emerged in 1965, when scanning electron microscopy (SEM), a now-ubiquitous imaging technique, first became available to university research labs. For most of human history, we could only measure what we could see with our eyes, so anything smaller than about 40 micrometres (0.04mm) was totally invisible. Optical microscopes, which rely on light and a series of lenses, allow us to see much smaller objects, down to 200 nanometres (that's 0.0002mm). This limit - known as the diffraction limit - is set by light itself. Similar to the way you can't accurately measure something smaller than the gap between the markings on a ruler, these microscopes can't resolve objects that are smaller than half the wavelength of violet light. Electrons, though, have a wavelength about a thousand times smaller than that, and because they're electrically charged, they can be collected into a focused beam. To take an image with electrons, we scan the beam over an object mounted inside a vacuum chamber. The electrons interact with the object, producing signals that are captured by a series of detectors. As the beam moves back and forth, it builds an incredibly detailed image of the object.*

Just like the first optical microscope, the SEM unveiled a previously unseen world to scientists, including Rodolfo Ruibal and Valerie Ernst at the University of California. They had long been fascinated by lizard feet, and suspected that the tiny, hair-like setae might hold the key to understanding their climbing ability. So they put small skin samples from the toes of a Tokay Gecko into an SEM to see what they could find. They first measured the setae, which varied in length between 30nm and 130nm, around the same size range as pollen grains. We'd later discover that there are about a million setae on each foot. But the added seeing power of the SEM unveiled something else – every single seta had a very bad case of split ends, branching into hundreds of even smaller hairs, which Ruibal and Ernst dubbed '**spatulae**'. These split ends were unfathomably tiny, right at the diffraction limit of optical microscopy. With their discovery came the complete picture of the gecko foot; it is a complex, hierarchical structure, with features of different sizes. The largest are the lamellae, the scale-like flaps of skin that cover each toe. On each lamella is a dense forest of setae; slightly curved, microscopic hairs. And at the tip of each seta are the flat, numerous spatulae.

---

* The electron microscope was invented in 1931. Back then, it was rather impractical and could only reach a resolution of 50nm. By 1965, many commercial systems were capable of achieving 1nm resolution.

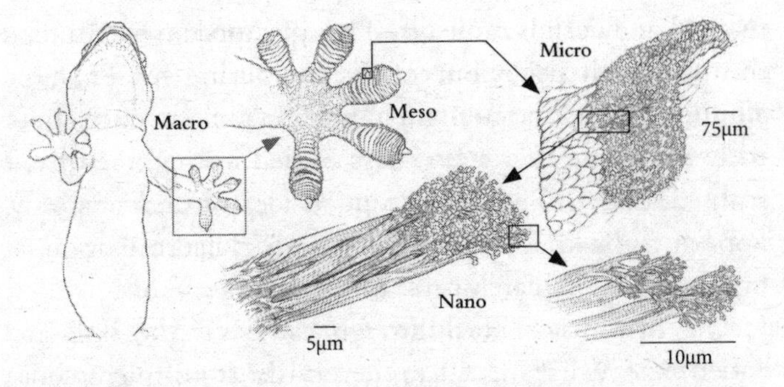

Figure 6: Gecko feet are covered in a range of complex structures of different sizes. Together, they provide the animal with all the tools it needs to climb any solid surface.

Just a year earlier, a zoologist called Paul Maderson had established that setae (and therefore spatulae) are made from β-keratins (beta-keratins). These are the much stiffer, more rigid forms of α-keratin (alpha-keratin), the protein that makes up the nails and hair of mammals. β-keratin is not an obvious choice for an adhesive, because it is incredibly smooth, slippery and hard. So Maderson concluded that the gecko's mystery adhesive mechanism couldn't be down to materials chemistry. He suggested instead that it must be a physical mechanism: one that depends on the huge surface area created by these hairs-on-hairs-on-scales. This led Ruibal and Ernst to come back to the idea that a friction-like force might be the culprit, and for the most part, that was where things stood for 30 years.

Fast-forward to the late 1990s. Kellar Autumn was then a post-doc, working on a research project for the US Navy. He and his colleagues were trying to develop highly mobile legged robots that could move around on rocky surfaces, and had initially looked to cockroaches as examples. 'We realised pretty quickly, though, that the problem was climbing,' he said. 'So we started looking for other animal models, and geckos came up.' Though Autumn had studied the nocturnal behaviour of geckos during his PhD, he admitted to me that back then, he didn't know much about their feet. 'But after a few days buried in the literature, I realised that, while we had a good idea of their anatomy, nobody really knew how geckos climb.' That realisation set him on a new research path; one that he's still on.

The first major breakthrough came via the skill and patience of Wai Pang Chan, a very dextrous microscopist then based at Berkeley. He managed to remove a single seta – just four-thousandths of a millimetre wide – from the foot of a Tokay Gecko, and carefully attach it to a pin. By isolating a seta in this way, Chan, Autumn and their colleagues could then measure the force needed to pull it from a surface; they could determine how sticky one seta is. From that, they could estimate the strength of gecko grip for a whole animal. The maximum adhesive force they measured was enormous – 10 times higher than had been predicted by any model. But even more interestingly, they found that gecko feet are, by default, non-sticky. These remarkable lizards switch on their stickiness only when they need it, and they do it through careful placement of their toes.

Place your hand, palm-down, on the table in front of you. Now slowly lift it. What happens to your fingers? If you're anything like me, they droop, or curl slightly towards your palm. But as I saw for myself when visiting the reptile experts at Wellington Zoo, climbing geckos do something else entirely. Each time they want to take a step, they first peel the tips of their toes backwards, curling them up away from the surface, and then lift their foot to complete the detachment. They go through the same process in reverse each time they plant a foot – the sole is placed down first, followed by a careful uncurling of their toes. What the Berkeley team realised was that this curling-uncurling action was the key to controlling the foot's stickiness, because it altered the angle between the setae and the surface. 'In our experiment, bringing the seta directly into contact with the sensor did nothing – it was totally non-sticky,' said Autumn. 'But when we carefully dragged it down, parallel to the sensor, we started measuring these huge forces. In fact, the faster we slid it, the stickier it got!' This really surprised the researchers. Usually when something starts to slip, it only slips faster. But the opposite was true for the gecko.

It works like this. The lightly curved setae on a gecko's toes usually curve proximally, pointing back towards the lizard's body. But, as they uncurl their toes in order to plant their foot and scale a vertical surface, gecko setae end up pointing in the opposite direction, forward, towards its claws. A very slight downwards slip of the foot (technically called a shear force) causes the setae to splay out, engaging the nanoscopic spatulae on their tips, and hugely increasing the foot's surface area – in this configuration, gecko feet are sticky. As the gecko climbs, its bodyweight has an effect, too. With gravity tugging on it, the setae on its toes make even more intimate contact with the wall, and this further increases the stickiness.

In other words, the gecko actually needs a little bit of slippage in order to reach its full stickiness. And when it does, the results are staggering. Most Tokay Geckos weigh in at somewhere between 200g and 400g (0.5lb–1lb). But, in theory, with all its feet and setae perfectly engaged, a single lizard could support a mass of 133 kilograms (293lb). The directional nature of gecko adhesion also means that the lizards pull harder with their front limbs, which makes them slightly larger than their hind limbs. This is in contrast to human climbers, who get most of their power from pushing upwards with their legs.

For a gecko, switching from 'sticky' to 'not sticky' is no big deal – it's just a matter of changing the setae angle, which it does by pushing its foot forward and curling its toes. Once these microscopic hairs reach an angle of 30° to the surface, they smoothly detach, releasing the foot. And it's here that the gecko shows its true superpower. The ability to stick is one thing, but a gecko can stick and unstick, again and again, on a range of surfaces, for the entirety of its life. Compare that to a strip of sticky tape – even if you do manage to peel it off a surface, you'd struggle to reuse it more than a couple of times. The speed with which geckos can detach their feet is also remarkable – the Tokay Gecko does it in just 15 milliseconds; blinking takes at least six times longer.* That ability allows the tiny Garnot's House Gecko (*Hemidactylus garnotii*) to climb walls at 77cm (30in) per second, which if we scaled it to the height of a human, would see the gecko nipping at the heels of Usain Bolt.

The outcome of all this is that gecko feet are the smartest on-off adhesive in the world. But I've deliberately omitted a very important detail – the actual mechanism by which billions of tiny hairs can make something sticky. For that, we need to zoom in beyond the setae, to the spatulae themselves.

Dipole

It turns out that these nanoscale hairs can make such intimate contact with surfaces because they tap into a vanishingly small force that acts between atoms. Called

* Blinking, according to Harvard, takes between 0.1 and 0.4 seconds, or 100–400 milliseconds.

van der Waals (vdW) forces after their discoverer, Dutch scientist Johannes Diderik van der Waals, they are not the result of chemical bonds formed between reactive molecules. Rather, they form between molecules that are already 'balanced'. To understand where they come from, cast your minds back to the picture of the atom you were presented in school science lessons.

You might now be thinking of negatively charged electrons, carefully arranged into concentric layers around a central (positively charged) nucleus. The reality is rather different. Electrons are constantly moving, whizzing around so quickly that, if we could see them, they'd look more like a blurry cloud than a collection of solid particles. This cloud is, on average, symmetrical, which means that the atom it's in has no overall electric charge. But at any given moment, it's likely that there will be slightly more electrons on one side of the atom than the other. This leads to a temporary, but very real, unbalance in the charges, known as an instantaneous dipole, where one side of the atom is negative while the other is positive.* This only really matters if another atom comes into close contact with the dipole. When that happens, the electrons in the new atom will rearrange themselves, so that it too becomes temporarily polarised, with its slightly positive side attracted to the slightly negative side of the original atom. This

* Some molecules, like water (H_2O), are what are called permanent dipoles, or polar molecules. Water's atoms are arranged into a pyramidal shape, with an oxygen at the top, and two hydrogens at the base. The oxygen end clings onto the electrons, making it permanently more negative than the hydrogen end. This allows strong bonds to form between water molecules, which is why water can remain liquid even at relatively high temperatures.

attracts more atoms, producing more temporary dipoles, and on and on it goes, resulting in a surprisingly stable system. The forces exerted between atoms (or molecules) via these weird fluctuations in electron position are what we know as vdW forces.

One major difference between these forces and general electrostatic interactions is scale. These vdW forces are much weaker than those that exist in traditional molecular bonds, where electrons are shared or donated and accepted. They also operate over tiny distances of about 10 nanometres. Once atoms are separated by more than that distance, they no longer experience vdW forces. Thankfully, our friend the gecko has all the hierarchical hardware needed to bridge that gap. The large, pliable lamellae on its feet help it to conform to surfaces, the densely packed setae provide a high surface area to further improve the contact, and the nanoscopic spatulae get so close to a surface that they can attract electrons in its individual atoms. So, gecko grip is really all about electricity.

Something that all geckos do, both in the wild and in captivity, is regularly change the orientation of their feet, depending on their activity. If a gecko is climbing up a wall, its four feet are all pointed in roughly the same direction – forwards, with their toes angling slightly away from their bodies (think 'jazz hands'). But if the gecko walks down a wall head-first, it turns its rear feet so that those toes point backwards, towards its tail. Now that we know that geckos stick via van der Waals, and that these forces are controlled by the orientation of setae and spatulae, changing their foot position makes a lot of sense. Gecko adhesion is entirely directional, so it only works when there are opposing forces acting on each other. As we've discovered, gravity helps geckos to climb walls – it pulls the setae downwards, switching on the stickiness by invoking vdW forces. But for a gecko descending a wall, gravity tries to reverse the direction of the setae, threatening to switch it into non-sticky mode. By turning its hind feet, the gecko can again make use of the 'good side' of gravity – the fully engaged setae on those two feet are more than enough to support its weight. For a gecko crossing a ceiling, it's all about balancing forces, too. Geckos position their legs so that their four feet splay outwards, radiating away from their bodies. This lets gravity tug uniformly on each foot, engaging as many setae as possible. Useful support in such a precarious position!

In their 2002 study, Kellar Autumn and his collaborators calculated that the van der Waals force between a single seta and a surface is around 0.04mN (= 0.00004N).* By any measure, this is a very small force. But when you remember that a Tokay Gecko has around four million setae to call upon, you quickly realise that it adds up to a sticking power that far surpasses what the gecko actually needs to support itself. It's often said that a gecko needs less than 1 per cent of its setae to be engaged in order to support its body weight. But as Alyssa Stark told me, this doesn't necessarily mean that they're over-engineered. 'You have to remember that all of these studies are carried out on carefully prepared, pristine surfaces in controlled labs,' she said. 'When I

---

* A Newton (symbol: N) is a unit of force. By definition, 1N is the force required to accelerate a mass of 1 kilogram at 1 metre per second squared ($1m/s^2$). Weight – a measure of the force of gravity on an object – is also measured in Newtons. So my smartphone, which weighs 145g on a kitchen scale, really weighs 1.42N. A millinewton (mN) is one thousandth of a Newton.

was out in Tahiti studying geckos, we saw them scale moss-covered trees, rough rocks and dirty, wet leaves. Wild geckos are very often missing toes, or have non-functioning ones, so they're unlikely to ever engage all of their setae at once.' So, it's less about overdoing it, and more about having precisely what they need to navigate changeable, complex environments.

Water

This led me down another path. Given that vdW forces rely on intimately close contact between surfaces, what happens in wet environments, like the Tahitian rainforests that Prof. Stark mentioned? Does the presence of water alter a gecko's ability to grip? Or, as I oh-so-professionally put it to her, will a wet gecko stick to stuff? Stark was the perfect person to talk to about this. Alongside her colleagues, she spent several years exploring the role of surface water in gecko adhesion, in order to better understand how it worked in real-world conditions. She started by measuring the adhesive force of Tokay Geckos on three samples of glass – dry, misted with water droplets, and fully submerged in water. Geckos were placed onto each surface, before being gently tugged backwards via a tiny motorised harness (yes, really), until all four of their feet moved. This let the researchers measure the force – known as the maximum shear adhesion force – required to overcome the gecko's stickiness.

They found that geckos can experience a significant drop in their adhesive performance on wet glass surfaces. 'This surprised us, especially as so many gecko species live in high-humidity, high-rainfall environments,' Stark said. 'We measured the lowest adhesive force when all feet were fully immersed, so water definitely interferes with the close contact needed for van der Waals-based adhesion.' But, she admitted, this situation is probably not all that common in the wild. 'Realistically, geckos are much more likely to interact with misted surfaces than go out in heavy rainfall and step into deep puddles.' Even so, Stark measured lower forces (or less stickiness) on misted surfaces than for dry-toed geckos walking on dry glass. In most cases, they'd still have *just* enough grip to support their body weight, but as the environment gets wetter, a gecko's adhesive force begins to fail. What's going on?

Back in Chapter 1, we learned that a liquid's ability to stick to a surface has a lot to do with surface energy, or wettability. Geckos' toe pads are superhydrophobic. They repel water so effectively that when the lizard puts its foot into a puddle, a tiny pocket of air forms around the toes; the water is pushed away, keeping the toes dry. But this water-shedding ability has limitations, determined by the surface the gecko eventually steps onto. In Stark's study, she focused on glass surfaces, which, because they are hydrophilic, attract water. When a gecko's foot comes into contact with the wet glass, it can't quite push all of that away and, as Stark explained, this interrupts the VdW forces that provide much of a gecko's grip. In addition, after 30 minutes of geckos having their feet submerged, their toes seem to temporarily lose their remarkable water-repellence. Water floods into the lamellae, reducing their grip even further, and making glass seem even more slippery.

But if the surface itself is hydrophobic, everything gets easier for the gecko. In that situation, both its feet and the surface repel water, and so the contact between them is effectively dry. That's the ideal situation for a gecko – with no water present, its setae and spatulae are all available for sticking. It also reflects the surroundings that many species encounter in the wild: from waxy leaves to tree trunks, hydrophobic surfaces are commonplace in nature. Importantly, geckos run a lot more frequently than they walk, and Stark later showed that this helps them to shed water from their toes more efficiently.

The realisation that wettability is a key player in gecko grip prompted several research groups to explore what happens to a gecko on engineered hydrophobic surfaces – most famously, the gecko versus Teflon match-up that was first discussed in the late 1960s. Experiments published by German scientist Uwe Hiller suggested that hydrophobic, low surface-energy materials like Teflon were just too slippery for geckos to climb. Even when he slightly increased Teflon's surface energy by blasting it with charged particles, his geckos still struggled to get very far. Experiments involving individual setae suggested the same outcome. So, it's perhaps understandable that Stark wasn't keen to test the material again in 2013. 'My undergrad student was super curious about what'd happen, though, so I relented.' What they found surprised everyone. According to their results, live geckos could cling to Teflon, but only when there was water present.

'It was one of those rare findings that both confused us, and confirmed what we'd seen in the wild,' Stark said. 'We know geckos can effortlessly scale the slipperiest of trees and plants even after a deluge, so water clearly isn't a serious problem for them. But our model simply didn't predict the outcome for Teflon.' The other results were less surprising – on materials with intermediate wettability, water didn't seem to make much difference. Geckos could cling just as well to both wet and dry surfaces. But superhydrophobic Teflon was the outlier – contrary to what we'd understood about van der Waals-based adhesion, water seemed to *improve* the gecko's adhesive performance.

The researchers suggested that this wasn't part of a wider trend, and was instead specific to Teflon. In the paper, they attributed it to Teflon's roughness. When dry, this roughness could cause air gaps that reduce the contact area between the surface and the gecko spatulae. When wet, the roughness may be somewhat smoothed out, allowing the toes to get into sufficiently close contact for vdW attraction. To be honest, I wasn't convinced by that explanation, and on the phone, Stark seemed to agree with me.

> We simply couldn't explain our results, or why Teflon was so different to other materials. In later work, we played around with its roughness and fluorination [a type of surface treatment] to see if anything changed. We found that the latter had more of an impact on adhesion. We suspect that electrostatics might be involved, but we don't know for sure.

There's seems to be no doubt that the dominant mechanism of gecko adhesion is van der Waals, but my conversations with researchers, combined with reading more papers than I care to admit, left me with the distinct impression that there's more to it. Despite constant and fairly intensive study, the gecko adhesive system may not yet have given up all its secrets.

For example, we still don't fully understand what happens to the keratin setae in wet environments. Human hair is incredibly susceptible to humidity, mainly because water helps temporary hydrogen bonds to form between neighbouring strands of the protein's α-form. Though there are some chemical differences between it and a gecko's β-keratin, it seems feasible that water could have an effect on its mechanical properties too. Autumn is in no doubt on that score. In a paper published in 2011, he found that the higher they raised the humidity, the softer a single seta became, but we don't know how that manifests on a 'whole animal' level. There are also a number of cell biologists who say that the keratin hairs have an additional function – the positive charges that naturally occur on the protein's surface might further enhance the vdW effect.

And finally, there's the 2011 discovery of some mysterious gecko footprints in a darkened research lab. 'We were not very popular when we published that paper,' laughed Stark when I asked her about it. 'Everyone says that the gecko uses a residue-free, clean adhesive system, but if that was the case, where did these footprints come from? They were leaving something behind, and we'd never seen that reported elsewhere.' Stark and her colleagues found that the residue contained lipids – compounds usually found in 'slippery' materials like waxes and oils. She also showed that these lipids were concentrated in and around the setae, which led her to suggest that it was linked to the keratin. But she admits that they can't yet explain why these lipids are present, or where exactly they come from. 'We just don't have those answers, though we suspect it has something to do with the constant trade-off between being sticky and moving fast. It might be that these lipids help keep the setae and spatulae clean and free of dirt. Or that the lipids might be doing something structurally for the hairs. Either way, it tells us that current models, based on homogenous pillars of β-keratin, are incomplete.'

These remaining questions only serve to make the gecko adhesive system more fascinating and worthy of study. Its performance has also made it a never-ending source of inspiration in the worlds of engineering and materials science.

Technology

Back in 2006, while working on a research project on water-repellent surfaces, I came across a paper that I'll never forget. Published several years previously in a scientific journal, it was written by a group from the University of Manchester. Two of the authors, Professors Andre Geim and Konstantin Novoselov, would go on to win the Nobel Prize in Physics in 2010 … though not for this particular piece of work.* What made this paper so memorable was a photo included on its final page – a familiar red and blue toy, suspended by one hand from a piece of glass. It was Spider-Man, made real.

Inspired by the climbing prowess of geckos, Geim and his colleagues had attempted to make a re-attachable dry adhesive, loosely based on the features of the lizard's feet. This was one of a flood of studies that had followed on from Kellar Autumn's 2000 paper. Like many of the other experiments, Geim's adhesive tape worked only in limited circumstances, and in truth, it wasn't even particularly gecko-like. Rather than having setae made from stiff, hydrophobic keratin, Geim's tape relied on flexible pillars of hydrophilic polyimide, which stuck to each other more effectively than they did to the target surface. So, though it worked initially, it lost its adhesive abilities after only a few attach-detach cycles. It really did make for a great photo, though.

That Nobel Prize-winning scientists failed to create a fully biomimetic gecko adhesive system shouldn't be a

* Geim and Novoselov's Nobel Prize was awarded in recognition of their 'ground-breaking experiments regarding the two-dimensional material graphene'. They were the first to isolate this unique material: a single sheet of carbon atoms that we'll revisit in Chapter 9.

surprise to anyone. Nature has had 200 million years to optimise the gecko's hierarchical adhesive system, whereas humans have only been trying to copy it for 20 years. As Mark Cutkosky, a Professor of Mechanical Engineering at Stanford told me, 'Whenever we examine biological systems in detail, we discover a daunting level of complexity, especially in locomotive systems. We can't get anywhere close to that.' And this is despite the ever-growing list of fabrication tools and processes available to scientists. But, Cutkosky continued, 'Maybe we don't have to reproduce everything. Maybe we can make a simplified structure that's good enough; an approximation of what we observe in nature.' This approach, of focusing on reproducing only the most important behaviours of biological systems, rather than attempting (and failing) to make perfect copies, seems to be working for Cutkosky. When I visited his Biomimetics and Dexterous Manipulation Lab, it was a roboticist's paradise: colourful plastic containers filled with electronics, tools scattered across the bench, and the rest of the space packed with robots and prototypes of all shapes and sizes – some designed to run over rough terrain, and others to fly and land on near-vertical surfaces. But it was the climbing systems that I was there to see.

Despite being a fan of the gecko's adhesive structures, Cutkosky was quick to say that it's not the only animal muse that he uses in his work. 'The first thing we think about is the application. What are you trying to do? What sort of surfaces are you likely to want to climb?' Once you've figured that out, he said, it's time to talk to the biologists. 'They help us identify specific animals that we can learn something from. We've repeatedly found that the most agile animals able to cope with a variety of surfaces are the ones with multiple adhesion mechanisms. Spiders, ants and cockroaches are all such multitaskers.'

In the case of the gecko, Cutkosky realised that while the hierarchical structures – the millimetre-scale lamellae, microscopic setae and nanoscopic spatulae – looked after most of the sticking, they didn't do it alone. The animal's incredibly flexible toes, as well as the claws present on some species, also played significant roles in how well the animal could cling to, and release itself from, surfaces. He explained, 'The combination of all of these is what lets the animal's feet conform intimately to surfaces at multiple length scales.' So, if they wanted to build a truly bio-inspired robot, they couldn't just focus on designing a material for its toes. They needed to look at the whole animal – 'to see the gecko as a system', said Cutkosky. Working with engineers, biologists (including Kellar Autumn), materials scientists from five universities and robotics company Boston Dynamics, Cutkosky's Stanford lab set to work.*

In 2007, the team introduced Stickybot, their first gecko-inspired robot, to the world. Weighing in at around 370g (13oz), Stickybot really did resemble a gecko. Its narrow body had a head, a tail and four feet which each had four soft toes. A series of motors moved the legs forward and back, up and down. A third set of motors allowed Stickybot's toes to do the characteristic curling-uncurling action that real geckos use to rapidly move across surfaces. The toes themselves were made from a polymer, arranged into strips and covered in thousands of

* Boston Dynamics is probably the most famous robotics company in the world today. Videos of their humanoid robots and four-legged robot dogs dancing to music regularly go viral online. I'll admit, I find them a bit creepy.

small wedges made from silicone rubber. It was, as Cutkosky said at the time, an 'analogous, albeit much less sophisticated' version of the gecko's hierarchical adhesive system, but it was effective. When shear forces were applied to the toes – for example, through the bot's weight tugging on them as it climbed – the wedges on its toes flexed, and made intimate contact with the surface. Though significantly larger than gecko setae, these bendy features still gave Stickybot access to the van der Waals forces that dominate real gecko adhesion.

Arguably, the coolest thing about Stickybot was the robotic system that made it move. As we've learned, a gecko can switch its stickiness on and off by altering the setae angle and balancing forces. If a gecko is running across the floor, it doesn't need to be sticky, so its setae lie flat. But once it begins to climb a wall, contact forces act on its feet, pulling from the palm out towards the toes. This spreads its setae outwards, making the feet super-sticky. All of these behaviours so instinctive to a gecko had to be designed into Stickybot. Forces were distributed via cable 'tendons'. A feedback system constantly monitored the position of its feet, and adjusted the applied force to attach or detach Stickybot's toes. The programmable motors managed 'leg phasing' – in other words, they made sure that Stickybot always had at least two opposing feet in contact with the wall at all times. Together, these features allowed Stickybot to climb smooth surfaces like glass, ceramic tiles and polished granite at 4cm (1.5in) per second; about one-twentieth the speed of a real gecko climbing a similar surface.

This robot had some other limitations. Thanks to the design of its ankle joints, Stickybot could only climb upwards. And despite being good enough to scale walls, the combination of force feedback and directional adhesive wasn't effective enough to manage ceilings. Later versions of Stickybot overcame many of these challenges, and perhaps more importantly, led Cutkosky and his colleagues, now scattered across different labs, to develop other technologies.

Take Assistant Professor Elliot Hawkes, from UC Santa Barbara. Back in 2014, he was one of Cutkosky's PhD students, working on a directional, reusable adhesive 'tape' inspired by the gecko foot. Hawkes was mostly interested in using the adhesive to support larger objects, like a human, to climb smooth, vertical surfaces. Because really, who hasn't dreamed of ascending a building like Spider-Man? He quickly realised that this wasn't going to be as simple as covering his body in gecko tape, because of something he called inefficient scaling. This is the observation that doubling the number of setae a gecko has doesn't actually double its stickiness - the real adhesive force tends to fall somewhat short of predictions. As Hawkes wrote at the time, 'Stickybot had an area of adhesive that should have supported 5kg [11lb] based on small-scale tests, but could only support 500g [1lb].'

Hawkes's solution focused on two main tasks:

1 getting the tape to conform to surfaces as intimately as possible, and
2 finding a way to distribute the load evenly.

The first was relatively easy to achieve – rather than using one large area of adhesive that could blend and flex unpredictably, he split it into many postage stamp-sized tiles. But sharing the load between those adhesive tiles so that they grip onto the surface at exactly the same time proved to be a lot more challenging. Drawing on the design of Stickybot's feet, Hawkes built an array of tendons to connect each of the adhesive tiles, but they had a secret. These tendons featured *degressive* springs, which, unlike normal springs, become softer the more they are stretched. This meant that the effect of even a small force was felt by all of the tiles at the same time, so the load was always evenly shared.

Hawkes used these ideas to design a climbing system based on two flat paddles connected to a stepping mechanism that applied the shear force necessary to engage the gecko-inspired adhesive.* As Cutkosky told me, 'It was unprecedented. Elliot could get nearly the same adhesive pressure [force per unit area] for an entire hand-sized paddle as for a single postage stamp-sized tile. It was this scaling efficiency that allowed him to climb.' And climb he did. Very slowly and very carefully, 3.7m (12ft) up a glass wall on the university campus. As far as I could find, development of the climbing system was largely handed over to DARPA (Defense Advanced Research Projects Agency), the US military agency that originally funded the work, and became part of a wider (and classified) research project into 'biologically inspired climbing aids' for humans. According to *Popular Mechanics*, a more recent DARPA version used a combination of directional adhesive tape and suction to cling to surfaces.

'Honestly, climbing's easy!' said Arul Suresh, a member of Cutkosky's team who joined our conversation by video

---

* There are lots of videos of Elliot Hawkes available online, including one where he calls out comedian and television host Stephen Colbert for his comments on Spider-Man not being plausible. ☺

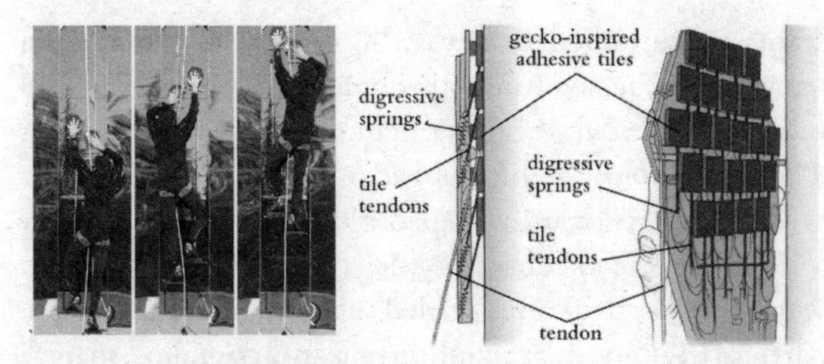

Figure 7: Elliot Hawkes successfully used his gecko-inspired adhesives to scale a building on the Stanford campus.

link from NASA's Jet Propulsion Lab (JPL). 'When you climb, all the forces point in one direction, so you can position the tiles to make use of that. But it gets a lot messier when the forces point in different directions. Have you picked up the ball yet?' The ball in question was sitting in front of me on the desk, its distinctive pointy-ended shape familiar to anyone who has ever watched a game of American Football (on purpose or by some horrible mistake. Guess which one I am).

Cutkosky reached over and handed me a strip of a grey rubbery material, with similar dimensions to a sticking plaster, but none of its obvious adhesive properties. It was suspended at its midway point from a piece of fishing line. 'This is the gecko tape,' he explained. 'Lower it down until it makes contact with the ball, and then lift.' I followed his instructions, and while I wasn't entirely surprised by the outcome, I was genuinely delighted when the ball rose majestically above the desk, picked up by no more than a bit of rubber on a string. 'This strip is actually made up of two patches that have been joined together, and they have opposing polarity – the wedges point in opposite directions,' said Cutkosky. 'One half wants to be pulled this way and the other half wants to be pulled that way. So when you put it on a curved surface, and pull through the centre, you load both of them in shear, switching on their stickiness. That's what lets you lift the ball.'

Grasping awkwardly shaped objects is something that a lot of robotic systems struggle with. Grippers based on soft, hollow structures called elastomer actuators are particularly good at conforming to complex shapes. They work by flowing a pressurised fluid – often air – through connected chambers arranged in an open ring. Positive pressure bends the actuators towards the object, which grips it. Sucking all the air out of the actuator causes it to peel back and disengage from the object. Elastomer actuators work well when the target object is fragile, but their low grasp forces and reliance on friction means that they are usually restricted to gripping objects smaller than themselves. Suresh and his colleagues realised that opposing patches of their gecko tape – similar to those that allowed me to pick up the football – could give soft grippers a way to handle larger objects. In 2017, they designed a system that combined the best of both. The hybrid device, which featured gecko tape on the gripping surface, could apply significantly stronger grip forces than traditional actuators, even when using much lower fluid pressures. It's similar to the difference between picking up a toy with a claw versus a dextrous human hand. The hybrid system was also much more forgiving of grasp location. Once both sides of the actuator could make contact with the object, it could generally pick it up. The 'high grip' version, which used three slightly wider actuators, could safely and repeatedly lift an 11.3kg (25lb) dumbbell, even when it was deliberately misaligned.

One of the other authors on that paper was Dr Aaron Parness, then a robotics engineer at NASA JPL.* Early into researching this chapter, I'd stumbled across a video of Parness. In it, he is testing a gecko-inspired gripper device on board the 'vomit comet'; the nickname for NASA's reduced-gravity aircraft. It uses a special flight trajectory to provide periods of near-zero gravity in which to test space-bound technologies. Parness can be seen manoeuvring a series of large objects – everything from a glass box to a cylindrical section of a fuel tank – with apparent ease. I was fascinated by the project, so I'd started digging into it.

It turned out that Parness had been heavily involved in the first Stickybot project, and had continued to work on gecko-inspired adhesives after joining JPL. But, despite the success of the gripper, the move into space technologies had come with its own challenges. 'I think I've had twelve flights on the vomit comet,' chuckled Parness, speaking to me via Skype. I was instantly envious, and my face clearly betrayed me, as Parness laughed again. 'Each time we've gone up, it's to test some aspect of a gripper or robotic system for use in space. We work on lots of different projects, but one thing that's important to all of them is gravity, or really, the lack of it.' And if you think back to how the gecko climbs, this makes sense. It relies on the force of gravity to load its setae and cause the nanoscopic spatulae to splay outwards. But, on an inverted horizontal surface like a ceiling, geckos have to use a special trick: 'They place their feet so that their adhesives are in opposition to one another, and then they squeeze,' explained Parness.

* Aaron left JPL in 2019 after nine years of working at the lab. At the time of writing, he was a principal research scientist in Amazon's robotics division.

'Geckos can do this between multiple legs and also a little bit between their toes.' Even in space, engineers wanting to emulate this effect have to rely on springs or cable tendons to pull opposing pads together and make them grip. 'In some systems, we've gone up to 28 tiles, all pulling centripetally towards the centre of a circle.'

The system that I'd watched Parness test in the vomit comet had used two different arrangements of gecko adhesive tiles – eight opposing pairs to grasp flat surfaces, and two pairs of curved grippers to grasp cylindrical or spherical objects. A network of pulleys could adjust the position of each of the tiles. To grip a free-floating object, the pulleys were tightened, and to release it, the tension was relaxed. Just like Elliot Hawkes's Spider-Man paddles, load-sharing was the key to achieving very low attachment and detachment forces that wouldn't simply push the target object away.

The gripper also had a 'nonlinear wrist' mechanism that acted as a cushion to absorb energy during impact. This helped the gripper to accurately grapple large objects on JPL's 'formation control testbed' at its laboratory in California. Affectionately known as the RoboDome, the testbed works like a giant air hockey table – large objects can move around without friction. 'It's a great way for us to play with relative motion and contact dynamics,' said Parness. In that test, the gecko gripper (mounted on one prototype spacecraft) successfully grappled and moved a solar panel (mounted on another spacecraft). That led to smaller versions of the grippers being tested by astronauts on board the International Space Station (ISS) in 2016, with another set of experiments carried out in May 2021.

'Our gecko grippers won't be the answer to everything, but they are very effective at moving things that otherwise don't have an easy place to grip onto,' explained Parness. 'That's very often the case with objects in space, so they could be a really useful tool for maintenance of the ISS or even grappling space junk.' The grippers are now being made by a commercial robotics company called OnRobot, and as well as finding use in space applications, they are regularly touted as an alternative to traditional vacuum grippers that are used to pick up large, smooth-surfaced objects here on Earth. But as Parness explained, there are some situations that gecko grippers struggle with. 'Dust can defeat the current version of our silicone gecko material. It's very easy to clean, but while it's dusty, it won't work as well. It also can't adhere to very rough surfaces.'

Recent research from Dr Amy Kyungwon Han, a postdoc in Cutkosky's Stanford lab, goes some way to addressing those limitations. In a paper published in late 2020, she described a hybrid gripper system that combines gecko tape and electrostatics. Rather than revisiting the question of whether real-life geckos utilise electrostatic attraction, Han adapted an existing gripping technology. Back in Chapter 1, we talked about how electrostatics can draw two dissimilar materials together. Well, in some cases, if a voltage is continuously applied to those materials, that attraction becomes more like a gentle clamping force that holds them together. This is the basis of electrostatic chucks: devices that are used throughout the semiconductor industry to move and manipulate completed devices and the fragile materials that form them. These chucks work in dusty environments, and they're significantly more tolerant of rough surfaces than gecko tape, which as we know, gets its sticking power from van der Waals forces. But their lifting capacity is fairly low, which means that unlike gecko tape, electrostatic chucks are typically only used to move lightweight objects.

Han and her colleagues set out to design a system that combined these technologies. They started with a wax mould of the same angled, microscopic wedges that are the basis of Stanford's gecko tape. This was first sprayed with PDMS (polydimethylsiloxane), a silicone-based polymer which thinly coated the entirety of the mould, while also filling the tips of each wedge-shaped hole. The rest of the wedges were filled with a different material – a rubber that surface charges are known to build up on. And finally, a film of yet another polymer was added, and this one houses the electrodes through which the voltage can be applied. These multi-layered adhesive pads were then applied to robot arms and used to pick up a range of different bulky items, including a bag of groceries, a glass container and a carton of canned drinks. On smooth surfaces like glass, electrostatics didn't confer any adhesive benefit. But on rough, porous materials like cardboard, the hybrid pads achieved an adhesion force that was three times higher than pads that didn't use electrostatics. And in all cases, the hybrid pads made it easier to lift things: the robot arms could apply just half the squeezing force to generate the same amount of lift. Though these pads are far from being commercialised, and there are some yet-to-be-answered questions around their robustness, the initial results are promising. Han said that they could find use in 'grippers, clutches, and other applications where good adhesion or friction is required', and because they're compact and lightweight and use very little power, 'they are suitable for small or mobile robots as well'. Han's postdoctoral work is partially supported by Samsung, while this project was funded by the Ford Motor Company – two industries that use robotic grippers on a daily basis. I'll be interested to see if this develops further.

The gecko will likely remain a source of inspiration for roboticists the world over. Its feet are the ultimate climbing and gripping tool, refined over millions of years of evolution, and capable of holding onto a diversity of surfaces through the manipulation of individual electrons. Their sticking ability comes from such a unique combination of effects that, if it didn't already exist in nature, I'm not sure anyone could have dreamed it up.

CHAPTER THREE

Gone Swimming

If you think back to the 2000 Olympics in Sydney, one name might come to mind: The Thorpedo. The 17-year-old Australian swimmer, also known as Ian Thorpe, was a sensation. Global audiences watched amazed as he won five medals (three gold, two silver) in just a few days. But Thorpe's medal haul wasn't the only topic of discussion around the pool that year, because those Games also marked the first wholesale appearance of a new generation of swimsuits. Extending from neck to ankle, with some versions also sporting full-length sleeves, the design was a significant departure from anything previously worn in an Olympic pool. At first glance, it had more in common with a wetsuit than the traditional suits or long shorts favoured by other competitors.

Made by Adidas™, Thorpe's bodysuit encased the swimmer in a lightweight, form-fitting shell of Teflon-coated Lycra. Years of research had gone into the suit's development. The compression granted by the fabric's tightness helped to smooth out some of the body's natural lumps and bumps, making the wearer more streamlined. And the ultra-slippery Teflon that covered almost all of the swimmer's body helped to reduce the friction they experienced in the water. The result, according to Adidas, was a demonstrable competitive edge – swimmers wearing their suits could move through the water more easily than those wearing other suits. News reports at the time mentioned what a swimmer was wearing almost as often as they talked about the competition results.

But Adidas certainly wasn't the only brand experimenting with high-tech fabrics; nor was it subject to the greatest amount of whispered controversy. In the same Games, swimmers like Inge de Bruijn (who won three gold medals and a silver) and Thorpe's teammate Michael Klim wore suits made by Speedo™. They were also made with a tight, stretchy fabric, and had bonded seams. But Speedo's textile included another, very different feature. It was covered in hundreds of shiny, ridged chevrons that all pointed towards the swimmer's toes.

This pattern gives a clue to the suit's origins. In 1996, during the Atlanta Games, a junior designer named Fiona Fairhurst visited London's Natural History Museum. She was doing her MSc at the time, and had convinced Speedo to take her on for a placement. She went to the museum mostly on the hunt for inspiration. 'I'd been starting to read about crazy stuff like biomechanical suits and biomimetic materials,' Fairhurst told me over the phone from the UK. 'As a swimmer myself, I kept thinking there must be a way to design our suits so that they could do more.' At the time, she said, there was an assumption that the key to making swimmers go faster was to create smooth surfaces. 'People always pointed to dolphins because they have slippery skin and are incredibly quick. But it's not a fair comparison. We humans are never going to have the dolphin's sleek, hydrodynamic shape. I went looking for examples that were a bit more like us: clunky in the water.'

That quest led Fairhurst to sharks, which in turn led her to Oliver Crimmen, a veteran museum curator and shark expert. On that visit, he introduced her to dermal denticles – the tiny chevron-shaped structures that cover a shark's skin, and which, Crimmen said, reduce drag. 'I was totally blown away by what Oliver showed me that day,' said Fairhurst. 'It made me wonder if we could create a fabric that would replicate the structure of shark skin. And maybe that would translate into improved swimming performance.' As a result of her research, Fairhurst was asked to lead the development of Speedo's next suit for elite swimmers. Project FASTSKIN was born, and it had a clear delivery target – the 2000 Olympics. 'My remit was literally just to make the fastest suit in the world, and for it to stand up academically,' she said.

The next four years of research took Fairhurst and her team all over the world, from Dunedin, New Zealand, to a fluid engineering lab in Nagasaki, Japan. The project transformed the way Speedo designed and tested swimsuits. 'They'd previously relied on results from wind tunnels,' chuckled Fairhurst. 'We started doing measurements in dedicated water flumes, where we could vary everything from the chemistry of the water to the air temperature.' They considered every aspect of the suit's fabric and construction too, 'everything from the seam position to the stretchiness of the thread'.

The resulting suit, described in one of Fairhurst's multiple patents, was made from single-layer, specially shaped panels 'of high stretch-constant polyester elastane fabric'. Swimmers could choose from a range of designs, from full-length leggings to sleeveless bodysuits, all made from this new fabric. Competitors wearing Speedo FASTSKIN went on to break 13 world records at the 2000 Games, and each further iteration of the suit design claimed new and ever more impressive functionality and performance. Speedo says that the version that appeared at the 2004 Games – the FASTSKIN FSII, also developed by Fairhurst – could reduce 'passive drag by up to 4 per cent'. It was this fabric that Michael Phelps was clad in when he won a staggering eight medals (six gold and two bronze) in the pool. And that caught everyone's attention, especially those in the scientific community.

If there's one thing that all scientists enjoy, it's testing claims made by big companies. Speedo don't make this stuff up – the company has a dedicated research facility in the UK called Aqualab, which employs good scientists and engineers, and Adidas is arguably even more research-heavy. The issue is that, because their focus is on developing commercial products, these companies don't often publish their research in traditional scientific journals. So as soon as the first bodysuits hit the headlines, researchers from other institutes started doing their own experiments on them. As a result, you can now find literally hundreds of peer-reviewed studies on the performance of 'shark-inspired' swimwear. Before we dig into a few of those papers, we need to get a better understanding of how swimmers actually move through the water, and the surface forces they have to overcome.

Swim

The first thing is that, regardless of what type of stroke they're doing, swimmers need to actively propel themselves if they want to keep moving. Yes, they can glide for a little while, but in order to keep their speed up, they must continuously push water out of the way using their arms and hands – these body parts generate 85–90 per cent of a swimmer's propulsion. The remaining 10–15 per cent comes from their legs and feet. As Fairhurst told me over the phone, 'a swimmer is only as fast as their levers'. Propelling themselves through the water takes a lot of strength, and it probably goes some way to explaining the typical V-shaped swimmer's physique seen at elite levels of the sport (wide shoulders and narrow hips).* The stronger a swimmer is, the more force they can apply to the water, and everything else being equal, the faster they'll be able to swim. So athletes work hard on improving their strength, endurance and what's called their peak power – the maximum force they can apply, multiplied by their velocity.

But these propulsive forces tell only part of the story. Swimmers also need to contend with hydrodynamic drag, the collective term for the resistive forces that act on individuals as they travel through the water. This loosely translates to the 'stickiness' of the water. The higher the drag forces, the harder it is for a swimmer to move forward, and there are three main types.

The first one is **form drag**, and as the name suggests, it is related to the shape and size of the swimmer. This is also the type of drag we experience when walking through the water. Because it scales as the velocity squared (v^2), form drag becomes increasingly important as swimming speed increases; a swimmer doubling their speed experiences four times as much form drag.

---

* Ian Thorpe has famously large feet. They're also, apparently, incredibly flexible. The *New York Times* reported that 'he can touch his shin with his toes', which is both fascinating and if I'm honest, a bit icky. The combination of flexibility and a large foot surface area are natural advantages, which likely contributed to Thorpe's swimming performance throughout his life.

Form drag is relatively easy to minimise in that a lot of the battle is won simply by being as streamlined as possible. This can be done via a swimmer's technique, posture and core strength. In a paper published in the *Journal of Turbulence* in 2001, the authors said that the general idea 'for most strokes is to have the shoulder/chest area create a gap in the water and the hips and legs (to) follow through that space. That usually translates into swimming with the body as level as possible.' There are some parts of the body where you want to maximise form drag, though – namely the forearms and hands, where that resistance helps in propulsion, so big hands are a plus. Something else worth noting is that form drag is slightly influenced by the density of the water, as swimmers tend to sit higher in salt water than in fresh. This buoyancy-related effect is something we'll come back to.

The second resistive force of interest is **wave drag**, which is the result of waves and wakes in the water. In a pool, these disturbances are created by swimmers themselves. As they swim at the surface and push water out of their way, they're continuously producing packets of water with different velocities. Swimmers effectively lose energy to this creation of accidental waves, which can seriously hinder their progress. In competition, lanes are used to keep swimmers separated, which minimises the impact that swimmers have on each other. Unfortunately, each individual still creates their own wave drag, and suffers the consequences of it.

Wave drag poses the biggest risk to performance because it dominates at higher speeds. It scales as the cube of the swimming velocity (v^3), which means that the faster a swimmer goes, the more dramatically wave drag increases

– for example, a doubling in speed would lead to eight (2^3) times more wave drag.* It is also amplified by vertical movement in the water. In an ideal world, a swimmer will direct all of their motion entirely along the horizontal. Unfortunately, that's not how humans are built. For one thing, swimmers need to occasionally raise their heads above the water to breathe, and as they kick their feet in the pool, their hips rotate. Given all that, it's not possible to remove wave drag entirely, but swimmers can minimise it by keeping their movements as smooth and steady as possible, and avoiding jerky, sudden transitions between strokes.

The third type of drag is **surface drag**, or **skin friction**. This sort of drag is affected by the surface roughness of a swimmer as they move through the water. It has a relatively minor effect on performance, because for a swimmer it scales linearly with velocity (v^1). It's worth briefly mentioning here though, because it relates to something we'll discuss much more in Chapter 4 – laminar versus turbulent flow. The way that a fluid, like water, moves across a surface can depend on how rough or smooth that surface is. What a swimmer really wants is a situation where they glide through smooth water with as little resistance as possible. But the presence of a bump or sharp edge on their bodies can cause the flow to become turbulent, creating tiny, swirling packets of water that absorb the swimmer's energy, robbing them of their speed. As a result, many competitive swimmers are obsessed with achieving smoothness, which might involve everything from

* Wave drag puts an effective speed limit on anything that moves through water near the surface – in boats, this is called the 'hull speed'.

removing body hair and exfoliating their skin, through to wearing swim caps and, of course, tight, high-tech swimwear.

The total resistance experienced by a swimmer is a combination of these three forces – form drag, wave drag and surface drag (skin friction). They all act continually on a swimmer, but the relative size of each force changes as that swimmer alters their body position and their speed. It's very difficult to isolate these forces and measure them independently, so instead, most swimming researchers and suit manufacturers talk about 'passive' and 'active' drag.

As you might guess, passive drag is the resistance that a swimmer experiences when they're not propelling themselves, so, when they're gliding or are in an unchanging position. Measurements of passive drag really only capture skin friction effects and some aspects of form drag, so they paint no more than a partial picture of the battle between propulsion and resistance. In reality, swimmers are constantly moving, and their velocity and effective size and shape change continuously throughout each stage of a stroke. If we want to get a more realistic estimate of the total drag they experience in the pool, we need to take force measurements while a swimmer actively swims.

Now, I say that as if it's a simple thing to do – but it's not. Even today, there's no standardised method for measuring active drag. This is likely to be because of how hard it is to design an experimental set-up that reflects what happens in a 'real' pool, without somehow interfering with the swimmer's ability to swim. But, especially for big corporations like Speedo and Adidas, it's worth the effort. Ultimately, if they can quantify these resistances, they may be able to design their suits, caps and goggles around them, helping their swimmers to become more slippery and, therefore, to move faster.

Measure

So, what are their drag-measurement options? Well, the technique of choice for several decades was called the MAD (Measuring Active Drag) system, and it was, indeed, mad.* Pioneered by Dutch researchers in the mid-1980s, it involved an air-filled PVC tube, submerged in a 25m-long (80ft) swimming pool. Fixed at regular intervals along that tube were paddles that an unnamed, male Olympic swimmer pushed against while doing the front crawl (or freestyle). These paddles were connected to sensors called strain gauges that measure force – in this case, the propulsive force created by the swimmer's arms. In order to turn that measurement into an understanding of drag forces, researchers had to make a fairly significant assumption: that the swimmer was moving at a constant speed. If that were the case, they could say that the forces must all balance out. In other words, the average propulsive force measured by the sensors would be equal to the average active drag. By measuring one, you could infer the other.

While the physics of this is totally valid, it leads to a pretty artificial situation for the swimmer. For a start, to keep all the measured forces pointing in one direction, they can't use their feet at all. In the MAD system, the swimmer's legs are tied together with a buoyancy aid, to help them maintain the ideal horizontal position. They also

* I say this in the Irish sense, where 'mad' means surprising or weird.

need to keep the length of their stroke constant, and hit every paddle accurately, so they can't swim as freely as they typically would. And, in later versions of the set-up, swimmers wore a snorkel to eliminate the need to raise their head above the water to breathe. So it was some way off reflecting reality, but that's not to say it wasn't useful. On the contrary, it seems to have been the first truly practical way to measure some of the fundamental forces of swimming, and it was widely adopted by researchers and coaches alike.

Since then, there have been lots of other recognised techniques. The nattily titled Velocity Perturbation Method (VPM) compared a swimmer's maximum speed when they swam freely to when they swam while being tugged backwards, with a known force, via a cable and waist-belt. The researchers' assumption was that the swimmer could generate constant power in both cases, so by comparing the two velocities, they could determine active drag. The Assisted Tow Method (ATM) was somewhat similar to VPM, but this time the swimmer was towed forward; they were assisted rather than resisted. While this allowed them to move more naturally, it also involved assumptions that may or may not be accurate.

I think my favourite technique is also the most recent one I came across, which was presented by Japanese researchers in 2018. In it, swimmers are placed in a water flume, and connected to a cable, as before. But this time, the cable is connected to two sets of sensors: one measuring forward swimming forces and the other the backward towing forces. Under their caps, each swimmer wears an underwater metronome – a waterproof timer that beeps at a fixed frequency – which gives them a target stroke rate to hit, even as the water flow speed in the flume is increased by the researchers. This method tends to measure higher values of active drag than the others. Does that mean it's any more accurate or more reliable? Honestly, I can't say for sure.

And that's the key thing to take away from all this. Even among the world's swimming experts, there's a lot of debate on exactly how to measure the forces that define the slipperiness of a swimmer in a pool. And if we can't agree on something as straightforward as that, how can we – or any sportswear brand – possibly quantify the impact of wearing a shiny suit?

There's no detailed, publicly available information on the specific measurement systems used by researchers at Adidas and Speedo. All we really have are their patents, their published 'drag reduction' metrics (for example, the 2003 Adidas JETCONCEPT suit claimed 'up to a 3 % improvement in swimming performance') and of course, their medal-winning results. For hard data on these suits, we need to look to studies from independent labs. Many have put them through their paces, all with the aim of answering the question, 'Do these suits really help people swim faster?'

Huub Toussaint, inventor of the MAD system, was among the first to carry out a controlled comparison between conventional swimwear and Speedo's full-body FASTSKIN suit. In his experiment, published in 2002, a group of thirteen professional swimmers (six men and seven women) swam a front crawl at specific speeds in a pool, hitting all of the necessary force-sensing paddles. They did this twice – once while wearing the FASTSKIN suit, and a second time wearing their usual, standard swimwear. This allowed Toussaint to directly compare the active drag experienced by swimmers, with and without the suit. While the suit seemed to confer a small advantage to some of the participants, taken over the whole group the average drag reduction was just 2 per cent. Toussaint concluded that the results 'do not corroborate the 7.5 % drag reduction claimed by Speedo as the benefit from wearing the Fast-skin. No statistically significant reduction in drag was found.'

Also in 2002, a group of Australian researchers put nine elite swimmers (five men, four women) into a pool, connected them to a simple towing mechanism, and measured the resistive forces they experienced as they moved through the water at fixed velocities. The researchers wanted to determine the effect of the Speedo FASTSKIN suits on both active and passive drag, when swimming at the surface of the water, and just below it. For the passive tests, swimmers were simply towed along while maintaining a static, streamlined position, with arms and legs fully extended. For the active test, they were towed while performing a flutter kick. Swimmers wore full-body FASTSKIN suits for half of the experiment, and their standard swimming costumes for the other half.

These scientists found that the suit had a significant impact on passive drag. Taken across all participants, it was lower at all speeds and at both depths. Lead author Dr Nat Benjanuvatra wrote, 'Results from the current study concur with Speedo (2000), in that the full length Fastskin™ swimsuit reduced passive net drag values … by a mean of 7.7 %.' The results were a bit more mixed for active drag, but the authors still concluded that FASTSKIN suits really offered lower frictional resistance than the standard suits, allowing swimmers to glide faster.

Indiana University's Professor Joel Stager took a less direct approach to the problem. Rather than trying to measure the forces applied by (or acting on) swimmers, he looked solely at swimming speeds. Stager took data from every US Olympic Trial since 1968, and used it to predict the times that swimmers were likely to achieve in the 2000 trials. He then compared this to the actual results from those trials when, for the first time, all swimmers were issued with bodysuits from several manufacturers. Stager's logic was that if these suits really did reduce drag, he should see a demonstrable change in swimmer speed, something beyond what could be achieved through training and nutrition.

Stager has said, 'We ran the statistics on their times, and it showed that the new suits made no difference at all.' In the paper – though I could find only a summary of the results rather than the raw data itself – Stager reported that just two events differed from predictions: the women's 200m backstroke, which was slower than predicted, and their 100m breaststroke, which was faster. No significant differences were seen in any of the men's events. At the time, Dr Brent Rushall – a leading sport psychologist and occasional collaborator of Stager's – wrote, 'Evidence of any performance benefit, as marketed by swimsuit manufacturers, did not exist at the US Trials. Manufacturers' claims should now be questioned.'

I could keep pulling out papers all day – there are hundreds to choose from – but even these three paint a picture of the problem. If you looked solely at the number of medals won by swimmers wearing bodysuits, you'd be hard-pressed to argue that they had no effect on swimming performance. But with so little scientific agreement on exactly how they worked, for many years high-tech swimsuits occupied a nebulous zone, sitting somewhere between marketing hype and technological breakthrough.

They also caused huge amounts of controversy, because many in the world of elite swimming saw these suits as a piece of equipment rather than clothing. Competitive swimmers have always taken steps to make themselves more slippery, including removing all body hair before a race, but these suits seemed to take it too far. As a result, they were accused of threatening the 'purity' of swimming as a sport. Six months before the 2000 Sydney Olympics, Brent Rushall submitted a strongly worded letter to FINA (Fédération internationale de natation, or the International Swimming Federation). In it, he argued that 'the sport could be irreparably changed' if bodysuits were formally adopted, and called for them to be banned. Rushall's argument centred on a specific FINA rule which, at the time, disallowed the use of 'any device or swimsuit that may aid his* speed, buoyancy or endurance during a competition (such as webbed gloves, flippers, fins, etc.)'. As Rushall saw it, it didn't actually matter if the suits worked or not – it was the intent of the suits that he took issue with. He wrote, 'The fact that the bodysuit manufacturers have publicly espoused the performance-enhancing quality of their products should be sufficient to make them unacceptable under this rule since, at worst, they may improve performance (which in swimming is measured by speed).' I think it's safe to say that Brent was not a fan of the Adidas and Speedo suits. And he wasn't alone, either.

---

* Ahhhh, the default male. Thankfully, this rule has since been updated and now refers to 'his/her'.

Ultimately though, as we know, the suits were approved for use at the Sydney Games and beyond. But the hostility around their use – and continued development – never really went away. Thankfully for us, nor did the scientific curiosity that surrounded them, so these days, we know a lot more about the workings of high-tech swimwear.

Shark

The marketing and imagery that surrounded Speedo's launch of both the FASTSKIN and its 2004 successor the FSII were very shark-heavy. But if you look at the patent granted to Fairhurst and her colleagues at the time, you'll struggle to find any mention of their fish muse. Fiona told me that, contrary to the hype, the FASTSKIN suit was never meant to be a direct, literal translation of a shark's skin. Her visit to the Natural History Museum 'was a catalyst, a starting point. It introduced me to dermal denticles – tiny, grooved features on a shark that manipulate fluid flow. Obviously, human beings are not sharks, but it got us thinking.'

So perhaps it shouldn't have been much of a surprise when, in 2012, a scientist called George Lauder declared that the textile fabric is 'nothing like shark skin at all'. As a Professor of Ichthyology, Lauder is much more interested in fish than in competitive swimmers, so he didn't specifically set out to test bodysuit fabric. Instead, he designed an experiment to look at the hydrodynamic properties of different types of real shark skin. The addition of man-made materials to the experiment was just a bonus. Lauder has been quoted as saying 'the literature on shark skin needed an upgrade. Once we got going, I thought it would be fun to look at the Speedo materials, because we

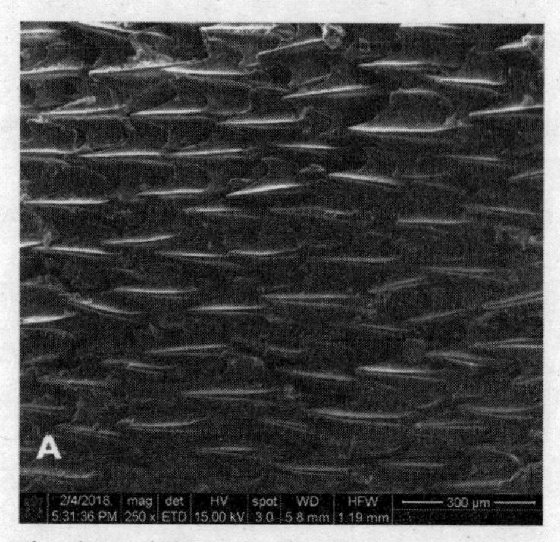

Figure 8: The dermal denticles from adult dogfish shark.

don't have a lot of quantitative information on the effect of surface structure.'

At this point, we need to take a brief interlude to talk about dermal denticles. These are the tiny, tough, tightly packed, tooth-like scales found on the skin of almost all shark species. I say 'tooth-like' because they actually have a lot in common with vertebrate teeth. My friend, shark scientist Melissa Cristina Márquez, told me that they 'are made from outer layers of dentine and enamel which surround a pulp cavity'. Denticle shape varies both within and between species of shark. And on any individual animal, you'll see different sizes of denticle on different parts of the body. Researchers have suggested that these structures perform multiple functions, from acting as armour to aiding in feeding.

What's most interesting about dermal denticles is that they are always aligned in the direction of fluid flow. If you ever find yourself up-close and personal with a shark – ideally in a hands-on museum, rather than the ocean – I'd suggest running your hand along its body from nose to tail, and then reversing the direction. It will feel smooth in the first case and rough in the second. That's all down to the directionality of the dermal denticles.

Interestingly, not all of the ocean's speedy swimmers have these features on their skin. Dolphins, by contrast, are remarkably smooth, as Dylan Wainwright, from Lauder's research group, explained via email. 'Some odontocete* species are known to have skin ridges, but we found that these had little or no hydrodynamic benefit. We think that – at least among the species we sampled – dolphins overcome friction by making their skin incredibly smooth and very taut.'† Could dolphins have been the inspiration for Thorpe's first Adidas suit? I couldn't find a reliable answer either way.

But let's get back to Lauder's shark experiment. The first step was to take samples of skin from two sharks known for their speed – the Shortfin Mako (*Isurus oxyrinchus*) and the Porbeagle (*Lamna nasus*). These were assembled into foils or flaps that, when attached to a robotic system, could 'swim' in a water tank. Half of these foils were rigid, made by gluing the skin samples directly onto solid plates. The other half were flexible, skin-only membranes. On some, Lauder deliberately removed the denticles using sandpaper to further isolate their role in drag reduction. The team used samples of FASTSKIN fabric to make a similar set of foils, and set to work comparing their underwater performance.

---

* Odontocetes are toothed whales, such as dolphins, porpoises and sperm whales.

† I love a good paper title, and this one is up there: Wainright, D.K. et al. How smooth is a dolphin? The ridged skin of odontocetes. 2019. *Biology Letters* 15, 20190103.

A key goal of the study was to make sure that the motion of the foils accurately matched the real-life swimming action of sharks. So Lauder used observations of live animals to programme his 'robotic flapping foil device'. And he could visualise the flow of water over the foils using a technique called Digital Particle Image Velocimetry (DPIV), which involves adding millions of tiny, reflective glass beads to the water, and illuminating them with a laser. As these beads moved past and around the foils, they could be carefully tracked with a high-speed camera.

Lauder found that on the flexible shark-skin samples, a spiral-shaped, low-pressure zone of water (a vortex) always immediately formed at the front of the foil. As the swimming cycle progressed, this vortex stayed attached to the foil, but gradually moved backwards along its length, before eventually disconnecting from the foil and dissipating into the water behind it. Once the foil returned to its starting position and resumed swimming, the vortex formed again, and the entire cycle repeated itself.* The DPIV tracking showed that within the vortex, the flow of water is temporarily reversed. The effect of this is that, as the foil swims, is it actively being 'sucked' forward by the water. No wonder sharks are so fast! The sanded shark skin samples didn't experience the same boosting effect. Although a leading-edge vortex formed, it detached from the foil very quickly, leading to swimming speeds that were 12 per cent slower.

These results led Lauder to conclude that denticles are the key way that sharks overcome active drag. But the effect was so pronounced that the denticles were doing more than

* If this is a little confusing, please stick with it – we're going to talk about vortex shedding a lot in the next chapter, I promise.

just reducing resistance – they seemed to actively increase swimming speed. Perhaps, as Lauder suggested to the *Harvard Gazette*, denticles on the mostly stationary shark's head are there to reduce drag, while those present on the ever-moving tail enhance thrust? We still don't yet have a complete answer to that.

But what about the FASTSKIN fabric? Well, the swimming performance depended on how the foil was moved by the robotic device. For two of the three motion programmes, the flexible foil actually swam faster with the suit fabric turned *inside out* than it did with the fabric in the correct orientation. And for the third, the swimming speed stayed the same no matter how the fabric was positioned. For Lauder, this showed that the surface patterns present on the surface of FASTSKIN suits have no effect on drag. In short, they do not act like denticles. He said, 'What we have shown conclusively is that the surface properties themselves, which the manufacturer has in the past claimed to be biomimetic, don't do anything for propulsion.'

While these results seem to draw a line under the shark-inspired, drag-reducing claims of FASTSKIN fabric, Lauder didn't entirely write off the performance of the bodysuits themselves. In fact, he said he's 'convinced they work, but it's not because of the surface'. So what other effects could be behind the record-breaking bodysuits?

Slick

There are lots of clues in a US patent issued to Fiona Fairhurst and her colleague Jane Cappaert in 2002. In it, the suit is described as 'a close-fitting garment', with 'panels of elastic stretch fabric joined at seams and shaped to conform with ... the abdominal region and ... the gluteal region'. From the sketches of the full-length bodysuit, there seem to be about 20 of these panels, all joined together using densely stitched, flat seams. The positioning of these seams is also carefully defined, and the patent says it's this that helps the suit achieve a 'highly-tensioned fit' over the body. The combination of these design features gives us a suit that is not only incredibly tight, but is also capable of supporting specific muscle groups.

Wearing such a suit, a swimmer would find it easier to maintain good posture and alignment as they swim. This not only frees up some energy that they can then put towards propulsion, but it also helps them to stay level, reducing form drag. Some researchers have suggested that neck-to-knee bodysuits could also reduce excessive hip movement in crawl, backstroke and butterfly, which would reduce wave drag. In addition, a group of Croatian biomechanists found that FASTSKIN-clad swimmers were less fatigued and displayed lower heart rates than those wearing standard suits. A later version of the Speedo suit, the FASTSKIN-3 (FS3), also claimed to improve the oxygen economy for swimmers by 11 per cent. Speaking about the FS3 to *Scientific American* in 2012, a Speedo research manager said, 'It's like miles per gallon in a car. You can swim at the same speed, but use less fuel. It allows a swimmer to go harder for longer.' So it's very likely that these suits provide an overall 'structural' support to swimmers' bodies, compressing their muscles, reducing fatigue and making them super-streamlined. This alone might be enough to account for their remarkable results.

Another word frequently mentioned around high-tech swimsuits is buoyancy. This is especially true for Speedo's 2008 Olympics suit, the LZR Racer, arguably the most controversial swimsuit ever made, but one that Fairhurst told me she 'had nothing to do with'.* The LZR took some lessons from the development of each of the FASTSKIN suits – its tightness compressed the swimmer's body into a sleek, streamlined tube and supported core muscles. But it went much further than that, largely thanks to its construction and the use of specialist fabrics. Unlike the suits that went before it, the LZR was not made from a knitted textile; instead, it featured polyurethane panels, separated by areas of elastane-nylon. Even the seams were different. Rather than using multiple threads to hold the suit together, the LZR's seams were ultrasonically welded: high-frequency sound was used to melt the two fabrics, forming a remarkably strong (and almost invisible) bond between them. NASA got in on the act too, working with Speedo to design a low-profile zipper that 'generated 8 % less drag in wind tunnel tests than a standard zipper'.

All this meant that the suit was almost impermeable; so repellent to water that it managed to trap pockets of air between the swimmer's skin and the inside of their suit. As a result, a swimmer wearing the LZR could float higher in the water, reducing the drag forces they experienced. The impact of these developments spoke for themselves. Speedo says that an astonishing 98 per cent of all available swimming medals at the Beijing Olympics were won by competitors wearing the LZR. That led an

* Having 'thought about nothing other than swimsuits' for eight years, designer Fiona Fairhurst left Speedo in 2004.

Italian swimming coach to compare the LZR to 'technological doping', and others agreed. Regardless, within months of the Beijing Games, multiple swimwear manufacturers launched their own versions of 'the rubber suit', including Arena's X-Glide and Tyr Sport's B8. These suits did away with the panels; made almost entirely from polyurethane, they made swimmers even more buoyant than their predecessors. Forty-three world records were set at the 2009 swimming world championships, which finally prompted FINA to step in.

Full-body polyurethane suits were banned in 2010. Since then, only permeable textiles have been allowed, and the suits have shrunk. As of 2017, for 'pool and open water swimming competitions with temperature above 18°C', men's suits must not extend above the waist or below the knee, and women's suits can only go from shoulder to knee. The technology development hasn't stalled, though. At this elite level of the sport, every millisecond counts, so manufacturers still search for speed gains through design … it's just no longer through the mechanism of skin-tight, buoyant, super-slippery bodysuits.

Float

Beyond humans, sharks and dolphins, there are lots of other things that have to deal with moving through water, the most obvious one (to me, at least) being boats. So, before we wrap up this chapter, let's talk briefly about hulls.

A boat's hull is its body. It's the part of the vessel that ploughs through the water, so as you might expect, boat hulls experience many of the same hydrodynamic forces as, say, a shark. No matter what form of propulsion a boat uses, form drag, wave drag and skin friction will all work to slow it down. Similarly, the total drag a boat experiences generally increases with speed. There are a few differences, though. For a start, boats are stiffer than sharks, and unless they're highly specialised, they don't change shape as they move. Boats also tend to operate near the surface of water, rather than being fully immersed in it. And unlike sharks, boats are in a never-ending battle against biofouling – the accumulation of algae, plants and animals (such as barnacles) that occurs on their hulls when they're stationary. All of these factors have an impact on a boat's performance, and on the particular drag mechanism that dominates. But if we wanted to make a boat faster, where would we start? The answer to that question depends on the boat, so to unpick it a bit, we can consider a few different designs.

First is the displacement hull, typified by canoes and fishing boats. As its name suggests, this type of hull works by displacing water, literally pushing it out of the way with its bulk. Because they rely on buoyancy, boats with displacement hulls can move without very much propulsion, and they're pretty stable. But their top speed is limited by how low they sit in the water – the greater the contact area between the boat and the water, the larger the wave it can potentially create. So if we want to increase the speed of a fishing boat, we need to reduce its wave drag. The most practical way to do this is to decrease its weight, which reduces the amount of water displaced, causing the boat to sit higher in the water.

Lightweight rowing shells, like those used in the famous annual boat race between Oxford and Cambridge Universities, also rely on displacement to stay afloat, but because they're so lightweight, very little of the hull sits below the waterline. They're also designed to be streamlined so they can slice through the water. Their long, thin shape makes them far less susceptible to wave and form drag. The dominant type of drag that acts on a rowing shell is skin friction, which, according to Oxford physicist and part-time rowing coach Anu Dudhia, accounts for around 80 per cent of the total resistance these boats experience. The only way to make racing shells faster (other than choosing stronger rowers) is to alter the surface finish.

If speed is what you're into, it's hard to look past the rarefied world of racing yachts, particularly those used in the America's Cup regatta. They can reach speeds in excess of 50 knots (92.6kph, or 57.5mph) despite not having an engine, and it's thanks to their ability to lift their hull out of the water and keep sailing. Two incredibly strong carbon fibre arms, called foils, protrude from the hull. At high sailing speeds, these foils are deployed below the waterline, allowing the hull to lift from the surface. These boats keep drag low by having as little of the boat in the water as possible.

Reducing hydrodynamic drag isn't merely an idle pursuit for wealthy people who race expensive boats. In the shipping industry, it is an obsession, because the easier it is for a powered boat to cut through the water, the less fuel it has to burn. And given that trade by sea accounts for almost 3 per cent of the world's carbon dioxide emissions, there's a growing incentive to make boats more efficient. Container ships are among the biggest vehicles ever built, but like wispy, low-slung rowing shells, they too have to fight against skin friction – at low speeds, it can represent up to 90 per cent of the total drag that acts on them. As we know from our swimsuit discussion, the skin friction between a solid and a fluid is strongly influenced by surface roughness, so shipping companies spend a lot of time and money ensuring that their boat hulls are as clean and smooth as possible. One of the ways they do this is via 'anti-fouling' paint, many of which contain poisonous compounds called biocides. Over time, these biocides slowly leach from the paint, preventing the growth of organisms that would otherwise significantly roughen the surface. As you might imagine, poorly managed anti-fouling paints can cause harm to the wider marine environment, so the regulations around their use are constantly being updated, and manufacturers are moving away from them.*

A spinout from Harvard University recently launched their contribution to this effort – a biocide-free paint based on the genus *Nepenthes*, a famously slippery carnivorous pitcher plant. While it might look smooth, the inner surface of *Nepenthes*' tube-shaped trap is covered in millions of microscopic pores, which are filled with water and/or nectar. This liquid acts as a continuously renewed layer of lubricant, and the effectiveness of the resulting surface is immediately apparent to any unfortunate insect that comes into contact with it. All that awaits them at the base of that slippery slope is a bath of digestive juices.

Professor Joanna Aizenberg and her colleagues at Harvard's Wyss Institute weren't interested in dissolving insects when they looked at this plant species. Rather, they used the concept behind its surface to design a whole host of

---

* Copper-based coatings are becoming popular among recreational boaters, but there is a lot of debate around their effectiveness and environmental impact.

ultra-smooth, ultra-slippery coatings that, by altering the lubricant, can repel everything from water and oil, to dust, bacteria and blood. They called this technology SLIPS, or slippery liquid-infused porous surfaces, and published their initial results in 2011. SLIPS was commercialised by Adaptive Surface Technologies, Inc., who have gone on to develop plastic containers that can be completely emptied and industrial tanks that don't hold on to any residue. Their anti-fouling marine paint became commercially available in 2019, and according to the product data sheets, it combines PDMS (a silicone-based polymer) and a compound widely used in hair conditioners. This combination would certainly make the paint super-slippery, and importantly, it won't leach anything into the aquatic environment - it just makes it harder for organisms to cling to the bottoms of boats. Multi-month field testing of the paint, carried out in 2020 in Singapore, showed that the coatings were able to 'largely deter settlement of marine mussels – one of the most invasive marine biofouling organisms – and to weaken their interfacial adhesion strength.'

There have also been multiple attempts to make boat hulls more slippery through controlling fluid flow with surface structures. Small, rib-like ridges called riblets, usually aligned parallel to the flow, have been investigated both experimentally and theoretically. And many of these studies found that on stiff objects such as boats, riblets do indeed reduce drag, by up to 7 per cent in some cases.

Professor Bodil Holst's approach involves using microscopic 'pancakes' to produce slippery surfaces. Initially developed on optically transparent materials like glass, Holst's surfaces repel *oil* by channelling *water* between the short, circular structures. 'The pancakes allow for a dynamic layer of water that flows stably on the surface,' she explained, speaking to me from her office at the University of Bergen. 'And because the water is present, and oils and water don't mix, the oils and other natural polymers common in marine organisms are prevented from sticking to the surface.' Holst described her decision to try this approach to self-cleaning as 'purely pragmatic. We felt that the pancakes would be easier to make than tall pillars, and like a short piece of chalk versus a long one, they'd be more robust and harder to break. The size of the flow effect was an added bonus.' In 2017, her team installed a pancake-structured window on an offshore water sensor that had typically been cleaned once a week. One year later, it was still spotless, which prompted Holst to file a patent application for these microstructures. Despite this, she remains cautious about the technology's potential. 'We still have lots of work to do. I'm not sure if biofouling will be the way this goes. The moment you include living things, everything becomes very unpredictable! We're exploring other avenues, like anti-icing surfaces – after all, this is Norway,' she laughed.

Another idea that's generating a lot of interest in the shipping industry is air lubrication – using a layer of bubbles to decrease the drag forces experienced by the hull. It works because air's viscosity, a measure of its stickiness, is a tiny fraction (about 1.6 per cent) of seawater's. So a ship could move through it more efficiently. The concept is not new; there are already several commercial air lubrication systems on the market, and they're being used on everything from bulk carriers to cruise ships. These existing systems differ slightly in their design, but they all rely on continuously blowing bubbles from the bottom of the boat. The longer the bubble layer remains attached to the hull, the greater the potential drag reduction. But keeping that

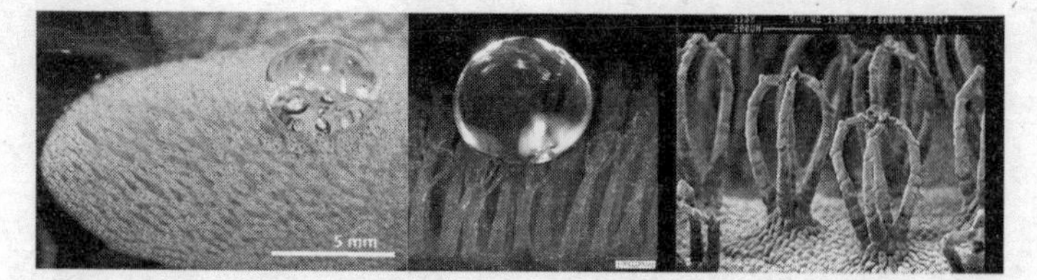

Figure 9: The Salvinia fern is incredibly water-repellent. Small, whisk-shaped structures on the tip of each surface hair give it this property.

air layer in place is a major challenge in the turbulent environment of the oceans.

As it turns out, there's a plant that might be able help with that. *Salvinia* is a type of fern that spends its life floating on water. Found mostly in tropical regions, this fern is rightly regarded as an invasive weed. It grows rapidly to form thick, extensive mats that can choke slow-moving waterways, and if unchecked, it can irreparably damage aquatic ecosystems. But, from a purely surface-science point of view, *Salvinia* is fascinating. I first heard about it via a radio interview with a physicist named Thomas Schimmel, and by the time he'd finished speaking, I'd already downloaded several of his papers. Back in 2010, Schimmel was working with Professor Wilhelm Barthlott, the botanist who first described the lotus effect (see Chapter 1). His research target this time was the *Salvinia* fern, and the team wanted to understand how the plant retained air so efficiently.

Salvinia leaves are covered in a dense forest of hairs, each about 2mm in length. If you zoom in a bit closer, you'll see that these hairs are rather complex – near the top, each one branches into four before meeting again at the top. Anyone familiar with baking will tell you that they look remarkably like teeny tiny whisks.

With the exception of the tip of each hair, *Salvinia* leaves are covered in wax crystals, which not only introduces a nanoscale roughness, but also acts to make the leaf superhydrophobic. By repelling water so strongly, the researchers reasoned, *Salvinia* was effectively creating its own cushion of air in and around its dense hair-whisk forest. That is what kept the leaves dry, even when they were submerged in water for weeks at a time. Barthlott, Schimmel and their colleagues set about replicating these features in a range of other materials. But each time, the layer of air would only stay in place for a few minutes. Time and again, their artificial leaves would emerge from the tank saturated with water. It was only then that they realised they'd missed something.

While most of the leaf worked to repel water, the non-waxy tips of each hair were working to attract it. These minute hydrophilic patches actively pinned water in place, trapping air underneath. As Schimmel explained during his interview, 'The hairs support a layer of air like the pillars of a tent support the volume of the tent … Air can't get out because the water is "glued" to the end of the hair, but the water can't penetrate further because the rest of hair repels it.' This double whammy, the balance between hydrophobic repulsion and hydrophilic attraction, is what stabilises the interface between water and air. It is the key to creating an air layer that sticks around permanently, rather than one that disappears after a few minutes.

The Salvinia Effect, as it's now known, could be used to make something almost unimaginable – boats that never get wet. Their permanent, protective coating of air would help to minimise biofouling when stationary, and reduce skin friction while in motion. Despite the potential speed gains, one area where we're unlikely to see air-trapping coatings is in yacht racing. As Emirates Team NZ's

Dan Bernasconi told me, friction-reducing coatings are strictly prohibited.*

It's possible, however, that such coatings will find a use in the wider world of boating: helping cruise liners, tankers and container ships to become a little more slippery. At this stage, it's hard to say how practical they'll be to implement. Barthlott and Schimmel are both still working on it, as are many other research groups across the globe. At the time of writing, Schimmel was collaborating with coatings specialist PPG on a large EU-funded research project called AIRCOAT. And in 2019, Barthlott published several papers on hydrophobic 'air retaining grids', which he says could make commercial air lubrication systems more efficient. Both ideas are robust, and the science is solid, but everything still seems to be at the research end of the 'research and development' scale.

One thing is for certain – the potential gains are huge. Ninety per cent of world trade happens by sea. If these or other new coatings could reduce drag, shipping companies would save money, millions of tonnes of fuel a year, and emit far fewer greenhouse gases. It's a win-win, right? Unfortunately, as with lots of conversations around climate change, the stumbling block might be ownership, rather than a lack of technical solutions. As two global studies experts wrote in *The Conversation* in 2018, 'national governments have largely ignored the carbon dioxide emissions from international shipping … This is a real problem because if no country is held responsible for emissions, no government will try to reduce them.' Individual companies such as Maersk

---

* The class rules for the 2020 boat specified just seven standard paint finishes that all competitors must use.

have pledged to slash their carbon emissions, but they haven't given any details as to how they're planning to do it. Speaking to CNN in 2018, Maersk Chief Operating Officer Soren Toft said they 'will need to find new technologies, new innovative ways of basically providing the future efficient ships'. Could super-slippery coatings be a possible answer to that call? I'll give you that one for free, shipping industry. Now please start talking to some scientists.

CHAPTER FOUR

Flying High

No matter where you go on Earth, you are blanketed in a fluid. Underwater, you are hyper-aware of this fact; even while wearing a full-coverage high-tech swimsuit, you can feel the interactions between your body and the water molecules that surround it. As we discovered in the previous chapter, some of the forces that water exerts on submerged objects are supportive and aid in flotation, while others resist motion, acting to slow even the fastest swimmers. Air can exert similar forces too, but they are far less obvious, especially if you're just an ordinary human wandering around on dry land. That's partly because air is a much, much less dense fluid than water – a litre of water contains about a thousand times as many molecules as a litre of air.* The fewer molecules a fluid has in a given volume, the lower resistance it offers, and the easier it is to push through. So in that sense, we can consider air to be more slippery than water. But there is still a cost to overcoming air resistance, and we pay for it with every movement. Each time you stand up from your chair, lift your arm or turn your head, you're barging aside billions of air molecules. Doing this requires energy.

---

* This factor of 1,000 comes from the values of density for air and water = 1.225kg/m^3 and 999kg/m^3 respectively (at 15°C and at sea level). You'll get the same result by comparing the number of molecules per litre, ~ 10^{25} in water vs. 10^{22} in air.

In a now-famous pair of studies from the 1970s, four men (one athlete and 'three non-athletes accustomed to prolonged physical exertion') undertook a series of walking and running trials on treadmills set up both inside a wind tunnel and out in the open-air. The aim of the research, which was led by physiologist and mountaineer Dr Griffith Pugh, was to better understand the relationship between air resistance and physical exertion.* He wanted to quantify the energy cost involved in pushing your way through air. Pugh did this indirectly, through measuring the amount of oxygen each participant consumed during their tasks. It turns out that regardless of whether you're walking or running, it really does require more effort to move through a headwind. One participant, walking at a constant speed of 4.5km/h (2.8mph), used almost three times as much oxygen when facing a strong wind (about 66.7km/h, 41.4mph) as when walking at that speed in calm conditions. For outdoor running without any wind, Pugh concluded that the energy cost of overcoming air resistance could be 8 per cent of a marathoner's total energy output, and up to 13 per cent for short-distance sprinters. While later studies would go on to downgrade those fractions – to 2 per cent for marathon running and 7.8 per cent for sprinting – the central point remains the same. Even in its sparseness, air exerts a resistive force on objects.

---

* Pugh lived a fascinating life. In 1953, he acted as the lead scientist on the first successful summit of Mount Everest. His research into the effects of cold and altitude on the human body would go on to transform high-altitude mountaineering, having an impact on everything from the design of suits and equipment to climbers' diets and fluid intake.

The effects of this force are both measurable and unavoidable, and they have shaped the evolution and design of much of the world around us. It's time to take a leap into aerodynamics.

Drag

Aerodynamic drag can occur anywhere that air is in physical contact with a solid body. There also needs to be a velocity difference between them. The object could be moving through still air (e.g. a ball thrown on a calm day), the air could be flowing past a stationary object (e.g. wind versus everything bolted to the ground), or both the air and the object could be in motion (e.g. driving into a headwind). It really doesn't matter which one's moving; as long as there's relative motion, there will be drag.

As you might expect, aerodynamic drag has a lot in common with the hydrodynamic drag that tugs on swimmers, sharks and boats. For a start, it tends to increase with speed; the faster you move through the air, the greater the resistance you face. It can also be split into various types, including form (or pressure) drag and skin friction. But there are a few more details that we need to add to our understanding of drag, and they're mostly related to the fluid. Let's start with **dynamic viscosity** (η, eta), which defines how resistant a fluid is to flowing. Really, it's a measure of the internal friction of a fluid, of how cohesive its molecules are. The higher its value, the less it likes to flow. We most often use viscosity to describe the stickiness of liquids. For example, olive oil is about 80 times more viscous than water. Further up the scale you'll find other liquids like honey (up to 10,000 times water) and ketchup (up to 20,000 times). Gases like air are also viscous, with values lower than those of liquids, though perhaps not as low as you'd think; there's a bigger difference between the viscosity of water and olive oil than between water and air.*

Where viscosity plays a major role in generating drag is right at the interface between the fluid and the solid. Imagine we have a box sitting on the ground in a stiff breeze. As the air buffets past, the molecules closest to the surface of the box stick to it, thanks to friction. As a result, their velocity is zero, just like that of the box. Molecules a little further from the surface are moving, but their speed is limited because they collide with these stationary molecules; think of it as jostling your way through a crowd of people who are all standing still – no matter how nimble you are, you will be slowed down. The further we move away from the surface (or the crowd), the less of an impact those stuck molecules (or people) have on the overall airflow. Eventually, once you get far enough away from the surface, you'll reach a point where all of the air molecules are moving at the same speed; this is called the free stream velocity. The layer in which the air velocity increases from zero to whatever the free stream value may be is called the **boundary layer**. And it is the behaviour of molecules within that layer that defines drag.

There are two main ways to describe this fluid flow. If the molecules are moving smoothly alongside and past each other in an orderly fashion without any mixing, that boundary layer is **laminar.** Laminar flow could be described as 'well-behaved', but the conditions in which it

* This is based on the value of both fluids at 15°C and at 1 atm of pressure 890 μPa.s (water) vs. 18 μPa.s (air) ~ 50x difference.

can occur are limited, which makes it unstable and relatively rare in nature. What you're much more likely to see is **turbulent** flow, with rapidly mixing swirls of fluid (called eddies). By comparison to laminar flow, turbulence is a chaotic, unpredictable mess … and I really do mean unpredictable. There's no fundamental mathematical theory that perfectly describes the properties of turbulent flow, and no universally agreed-upon definition of it.* Of course, that hasn't stopped scientists and engineers from studying it – it just has an impact on the tools they use. Typically, understanding turbulence involves statistics and probability, averaging behaviour over large groups of molecules. From a practical viewpoint, this approach has worked so well that it has allowed humans to design systems that travel at unimaginable speeds (more on that later) but mathematically, turbulence remains an incredibly complex topic.

Laminar and turbulent flow can (and do) co-exist, and the transition between these two states is a central tenet of fluid dynamics. For any given fluid flow, this tipping point can be predicted via the Reynolds number (*Re*), named for its nineteenth-century inventor, Osborne Reynolds. *Re* is the ratio of a flow's inertial forces (its tendency to keep moving) to viscous forces (its stickiness). It's usually written like this:

$$Re = \frac{\rho v L}{\eta}$$

* The Navier-Stokes equations, developed in the 1820s, are widely used to model fluid flow, but without a mathematical proof, we can't say that they will always work. There is a $1 million prize on offer to anyone who can develop one.

With ρ (rho) = density of the fluid, v = flow speed, L = distance or length, and η = dynamic viscosity of the fluid. This equation tells us that the Reynolds number is low when viscous forces dominate, which happens when the fluid has a high η, and/or it's moving slowly (low v). This leads to the smooth, sheet-like motion of laminar flow. If inertial forces dominate – for example, where the fluid has a high speed and/or a very low viscosity – the Reynolds number is higher, and turbulent flow results.

But this simple equation tells us something else important: if you take a fluid of any viscosity and accelerate it (increase v), that will drive up the value of *Re*. In other words, simply speeding up a fluid can be enough to tip it from laminar (low *Re*) to turbulent (high *Re*) flow. We see this all the time in the real world. Take a waterfall as an example – its upper reaches are laminar, with water flowing clear and smooth, but further down, as the stream is accelerated by the force of gravity, the disordered chaos of turbulence kicks in. Similarly, the smoke rising from a lit cigarette initially travels in a narrow, vertical path, before transitioning into chaotic mixing, a few centimetres above the tip.

This onset of turbulence in a fluid happens at different values of *Re*, depending on where the fluid is flowing. If it's in a narrow pipe, the boundary layer will change from laminar to turbulent when *Re* > 2900. For our box sitting in a breeze, *Re* might need to reach 50000 before it trips into turbulence. The beauty of the Reynolds number is that, despite its simplicity and huge range, it can be used to make predictions about fluid flow that apply at any scale. For example, this ratio helps us to understand how a full-sized aircraft will perform in flight, based on measurements taken on a miniature model of it.

Nowhere is the relationship between turbulence and drag more stomach-churningly evident than on a plane flying in bad weather. The bumps and bouncing that passengers experience reflect what's happening in the fluid. In a turbulent boundary layer, fluid motion is dynamic and unstable – as molecules mix and swirl, pockets of faster fluid are brought closer to the surface. These eddies cling to the surface, rapidly increasing the skin friction drag. This is why turbulent boundary layers are a nightmare for aerodynamicists – their presence results in significantly more drag than laminar layers. As we'll discover, the pursuit of making objects move faster through the air is really about keeping turbulence to a minimum, or at least delaying the laminar-to-turbulent transition until it's as far away from your precious surface as possible. Even the tiniest obstacle – a few specks of dirt or a scratch or dent – can be enough to transform a laminar flow into a turbulent one, greatly increasing the drag experienced by that object. Airlines and trucking companies keep their vehicles clean and smooth because the cost of not doing so can be seen in their fuel bills.

But as an English engineer proved at the turn of the twentieth century, when it comes to flight, roughness isn't always a bad thing.

Balls

Depending on which dubious internet list you believe, either all or a majority of the world's most viewed sports involve kicking, hitting, throwing or catching some sort of ball. And whether they realise it or not, skilled players of these sports have developed an instinctive and sophisticated understanding of aerodynamics; they know how to get the ball to do what they want. A bowler who rubs one side of a cricket ball on their leg before delivering it down the crease to the waiting batter. A grand slam winner positioning their racket to send a tennis ball *just* skimming over the net. Or a midfielder knowing the exact spot to strike a football to make it curl beyond the keeper's reach. In these moments, each player makes decisions that actively alter how the ball moves through the air. They beguile boundary layers and toy with turbulence, manipulating complex surface interactions to help them win. But they don't do it alone – the ball can also have a role to play in how it moves.

Take golf, a game that's been around since at least the 1400s, and which is now played by tens of millions of people every year. Until the turn of the twentieth century, most advances in golf had focused on the material of the ball. Thanks to its excellent flight characteristics, the featherie, a woven leather sphere densely stuffed with goose feathers, reigned supreme for more 300 years. But it was the move to the guttie, made from a rubber-like compound called gutta-percha, that first helped to popularise the sport.* Gutties were cheaper and easier to make than featheries, they were more robust, they didn't soak up moisture and they flew

* Gutta-percha has been widely used in various ways for centuries, but was first pioneered by Malay people. It is a sap extracted from the tree with which it shares a name. It's a thermoplastic, which means that it is very pliable above a certain temperature, but solid when it cools.

further. Whether the ball was leather or rubber, these early manufacturers were convinced of one thing – achieving a smooth surface was the key to long-distance ball flight. Players began to suspect otherwise, though, finding that old gutties covered in dings and dents from use often flew further than pristine ones. The effect was so noticeable that players started deliberately damaging their gutties. Manufacturers had to respond, and by 1890 textured golf balls were commercially available. An array of raised bumps that made the ball look a little like a blackberry (or bramble) was the most popular pattern, but there was very little solid science behind it.

Enter William Taylor, a precision engineer and recreational golfer from Leicester, England. Frustrated by the haphazard approach manufacturers were taking to golf-ball design, Taylor set out to do his own experiments. He started by building a mini wind tunnel: a glass-fronted chamber in which smoke could be blown at different speeds over patterned ball surfaces. By looking at the behaviour of the smoke along each surface, he could determine the optimal option for flight. His conclusion? A dimpled ball, one that inverted the bramble pattern, offered the best performance. In Taylor's patent, awarded in 1908, he writes that the dimples 'must be substantially circular in plan and substantially evenly distributed, they must be shallow, and their sides, particularly at the lip of the cavity, must be steep.' Today, all golf balls bear the mark of Taylor's influence, with each one featuring somewhere between 300 and 500 shallow dimples on its surface.

To understand how the presence of these dimples influences the aerodynamics of the ball, we first need to think about a smooth ball in flight. As we've learned,

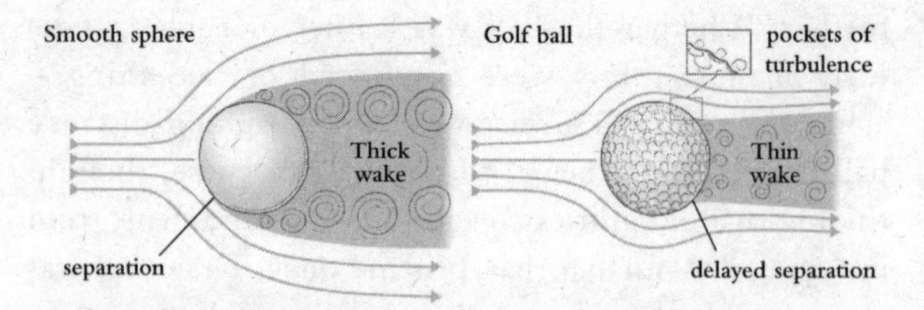

Figure 10: A smooth ball and a dimpled ball move through the air very differently. This is due to the size and shape of the areas of turbulence that build up on their outer surface.

a smooth surface can be helpful if you're interested in maintaining a laminar boundary layer. Just like with our static box in the wind, skin friction is present in laminar flow, but it's relatively low, meaning that drag should also be low. So far, so slippery. But skin friction is not the only resistive force that a moving ball experiences. Thanks to the billions of air molecules that bash into it as it flies along, there's also a build-up of pressure in front of the ball. In that high-pressure zone, some of the particles are stationary. The moving air, in an attempt to stay attached to the surface, follows its contours, and forms a very thin laminar boundary layer on the front half of the ball. But somewhere near the ball's poles, the situation changes. The thickening of the boundary layer combined with the accelerated air molecules causes the flow to suddenly detach from the surface, producing a wake – a turbulent, low pressure zone behind the ball that acts to slow it down.

This boundary layer separation is the source of **pressure** (or **form**) **drag**. Because it relates to the shape of an object, it can largely be ignored for flat surfaces like our box, but it is the dominant form of drag experienced by a ball in motion. The wake also helps to define the aerodynamic size of an object – the larger the low pressure zone, the bigger a ball 'looks' to the air, and the more significant the (pressure) drag.

Shallow dimples on a surface act like a series of tiny imperfections. At speeds typical of a golf ball, their presence has no real impact on the airflow at the front of the ball – it's largely laminar. But rather than detaching at the poles, the air molecules trip over the dimples, making the boundary layer turbulent. This increases the skin friction, helping little packets of air to cling to the surface for a little longer as it flows around the ball. The dimpled surface effectively delays the boundary layer separation, producing a much smaller wake and reducing the pressure drag significantly – more than enough to make up for the slight increase in skin friction.

When Taylor set up his smoke chamber, he was looking for the presence and location of these turbulent eddies. He realised that the further behind the ball they appeared, the lower the overall drag would be. The size, shape, depth and position of the dimples all have an impact, but in general a dimpled ball experiences about half as much drag as a smooth ball of the same size. And that means it can travel twice as far. All this goes to show that sometimes, a well-designed rough surface is more useful than a silky smooth one.

Sport gets even more aerodynamically interesting when the ball is also spinning as it travels through the air. Just like before, there's an area of high pressure at the front of the ball and an area of low pressure at the back, and a mixture of turbulent and laminar boundary layers on the surface. But when a rotational motion is added, the 'stuck' layer of air molecules is dragged along with the surface. So if the ball is thrown with backspin – rotating clockwise when viewed side on – the boundary layer at the top of the ball is moving in the same direction as the airflow. This allows the air to stay attached to the surface for longer, shifting the

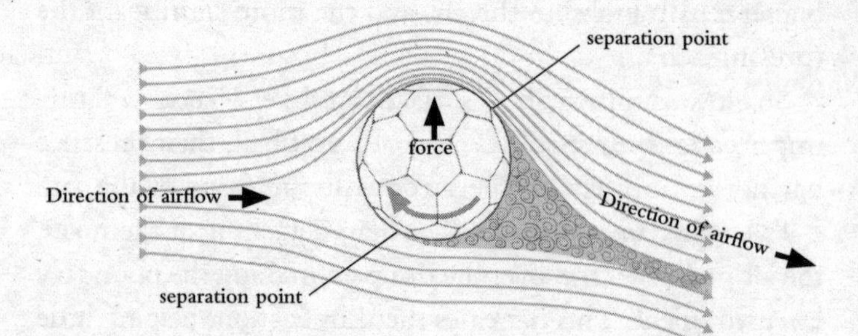

Figure 11: The Magnus Effect explains why a kicked football can 'curl'.

position of the flow separation towards the back of the ball. At the bottom, the boundary layer is moving in the opposite direction to the airflow, causing it to separate from the surface almost immediately, very close to the pole. The asymmetry in these separation points means that most of the airflow is deflected downwards. And because this exerts an equal but opposite force on the ball (thanks Newton!), the ball is deflected upwards. The faster the ball spins, the larger the force, and the bigger the deflection.

The result of this unbalanced airflow around an object is similar to the lift forces that birds and aircraft exploit in order to take to the skies (more on that later), but when the object is a spinning ball, it's referred to as the **Magnus effect**, and it appears in lots of sports. It's the origin of the 'bend' or curl that footballer David Beckham was famed for.* By striking the ball off-centre

* The Magnus effect was named after German physicist Heinrich Gustav Magnus, who formally described it in 1852. Beckham's ability to score from seemingly impossible free kicks by curling the ball around a wall of defenders was the inspiration for the 2002 movie, *Bend It Like Beckham*.

while leaning his body back, he could give the ball the spin needed to lift and curl high above the keeper and into the goal. The extreme version of this, known as the banana kick, has seen professional players score unassisted goals from the corner flag. In tennis, players often impart a ball with topspin, which rotates it in the opposite direction from backspin. That flips the resulting force, causing the ball to curve down towards the court. The true masters of spin are table tennis players, who can use their textured, rubber-coated bats to apply huge amounts of topspin, backspin or even sidespin to the lightweight ball.*

But there are two sports in which the Magnus effect gets a major upgrade. On the face of it, cricket and baseball might appear to be very different, but they do share some similarities. They're both bat-and-ball games, in which one team (the fielding side) delivers the ball to the opposition (the batting side) and at some point during the match, the sides switch roles. They have the same overall objective – to score as many runs as possible while minimising losses. And most importantly, for our purposes, they both use a ball with raised, stitched seams.

In a top-quality cricket ball, the seam consists of six rows of stitches which run around the equator of the ball, holding its leather hemispheres together. In contrast, a regulation baseball is constructed from two figure-of-eight-shaped pieces of leather, which are joined together by V-shaped stitches, producing a curving seam. In both cases, the seam is much rougher than the leather around it.

* Dianna, AKA the Physics Girl, has a lovely video about the reverse Magnus Effect on her YouTube channel. Search for "How smoothness of a soccer ball affects curve" to find it.

The presence of a seam gives a skilled bowler a lot of aerodynamic choices. By altering how they deliver the ball, in terms of speed, spin and its seam position, they can define how much of the airflow around it is turbulent or laminar. In cricket, one of the main specialisms is swing bowling. As the name suggests, the goal is to get the ball to swing, or deviate sideways, as it moves. To do this, a bowler needs to make the ball a little less symmetrical. Rather than holding it so that the seam is pointed straight towards the batsman, they angle it slightly, so that it points just left or right of centre.* From the air's point of view, the ball now has one smooth side (the leather) and one rough (the seam). As the ball flies through the air, the boundary layer on the rough side becomes turbulent as it trips over the seam. The air clings to the surface, only separating from it towards the rear of the ball. On the smooth side the flow remains laminar, but it separates from the surface closer to the front of the ball. An unbalanced airflow results, but unlike in the Magnus effect, this one has nothing to do with spin – it's all down to the asymmetry caused by the seam. And because it happens on the sides of the ball rather than its poles, the deflection it produces is also side to side. A brand new ball will always swing towards the direction that the seam is pointing. If a swing ball is also spinning when it's bowled, its trajectory can combine elements of both effects, making it even harder for the batsman to anticipate its path.

---

* The MCC – the 'custodians of the Laws of Cricket' – updated its laws in 2017 to make its language more gender-neutral. The term 'batsman' is still in use in both the men's and women's game. However, in 2021, popular cricket news website ESPNcricinfo announced that it will instead use 'batter' in its coverage. So, it's fine to use either.

Bowlers can accentuate the asymmetry of a cricket ball by regularly polishing one side of it using the fabric in their trousers or shirt, and this process becomes especially important later in the game. Frequent impacts with the ground and bat means that a cricket ball gradually roughens over time. In order to retain the ability to swing the ball, the bowlers need to carefully maintain the one smooth side; their sweat and saliva aid in the polishing. While this effort is seen as a central part of the game, any action to deliberately roughen or scratch the rough side is expressly outlawed (although that hasn't stopped some teams from attempting it).*

In baseball, the most interesting pitch is the knuckleball. Famously difficult to hit, it takes a relatively slow but erratic path from pitcher to batter that sometimes seems impossible. Really, it's all down to aerodynamics. Keeping spin to a minimum is the key to throwing an effective knuckleball. This might seem a little counter-intuitive, given that we know that lots of spin leads to a large Magnus effect, but a knuckleball isn't about producing the biggest deflection. It's about unpredictability. What a pitcher wants is for the ball to squirm and flutter through the air. Spin won't get you that, but a seam that changes position mid-flight will.

It starts with the grip. Typically, if you hold a baseball in your palm, your fingers briefly drag along its surface as you release it, causing it to rotate. This added backspin can contribute to lift, but for a knuckleball, you want to keep the number of rotations low. To do this, early adopters of the technique would clutch the ball between their thumb and their knuckles (hence the name) or hold the ball in

* In 2018, members of the Australian cricket team attempted to scuff the ball using sandpaper during a high-profile test match against South Africa. The captain and vice-captain were both implicated in the scandal, alongside a more junior player. All received suspensions.

their fingertips. Most current players dig their fingernails into the leather or the seam. According to Rod Cross, a physicist from the University of Sydney who has been studying ball aerodynamics for decades, a knuckleball typically rotates just one to three times on its way to the batter. As it does so, it continuously presents a slightly different seam pattern to the air, changing the proportion of the ball that appears rough or smooth. The mixture of laminar and turbulent airflow that then forms on the ball alters the size and direction of the forces acting on it. Over the distance of 60 feet (about 18m) between the pitcher's mound and the home plate, this can divert the ball from its expected path by tens of centimetres. Cross's collaborator, Alan M. Nathan from the University of Illinois, said that the erratic movement of a knuckleball, 'the deviation from a straight-line trajectory, is random from pitch to pitch', which suggests that even an expert knuckleballer can't predict exactly where the ball will end up once it has left their hand. What chance does the waiting batter have?

Skill and training allow top players in these sports to tap into a host of interesting aerodynamic effects. As a result, using nothing more than a club, racket, bat or basket, it's possible to propel a ball at speeds in excess of 300km/h (186mph) – that's faster than one of my favourite supercars, the Ferrari Monza SP1.* But that's still not fast enough for me.

---

* According to *Guinness World Records*, 'the fastest projectile speed in any moving ball game is c. 302km/h, 188mph', which was reached in a sport called jai-alai or zesta punta. It uses a long, pointed, curved basket, worn on the hand, to accelerate a small and incredibly hard ball against a wall. To date, the fastest ever golf drive reached 349.38km/h (217.1mph) – this happened on a dedicated driving range, rather than during a traditional game of golf. A Ferrari Monza SP1 has a top speed of 299km/h, or 186mph.

Mach

I was six years old when I first saw *The Right Stuff* – the epic film adaptation of Tom Wolfe's book – on TV. With a late evening start, I definitely wasn't the target audience, but I watched, transfixed, until I could no longer keep my eyes open. As my mum tucked me into bed, she hastily promised that I could finish watching the movie another time. Just a few weeks later, a VHS copy was waiting for me under the Christmas tree. I made the whole family watch it after dinner. This time though, I stayed awake until the final credits.

The Right Stuff covers a 15-year period of US history, when aviation first began to make way for astronautics. It's filled with names familiar to anyone interested in space exploration: John Glenn, Alan Shepard and the other *Mercury Seven* astronauts. But it was the other story – that of test pilots wrestling high-speed aircraft through the thin air of the Californian desert – that really stuck with me. I became obsessed with Chuck Yeager, who, with more than a bit of help from his flight engineer Jack Ridley, became the first human to break the sound barrier, on 14 October 1947.*

Yeager's bright orange *Bell X-1* was the first successful supersonic aircraft, but even to my six-year-old eyes, it bore only a passing resemblance to a commercial aeroplane. OK, it had wings, but they were short and razor-thin. It was powered by a rocket instead of a turbine engine. And it didn't take off from the ground either; rather, it was

---

* After a lifetime of pushing experimental aircraft to their absolute limits, Yeager passed away (aged 97!) on 7 December 2020, a few months after I finished writing this chapter. I never got to meet him, but I am lucky enough to have a signed copy of his autobiography. It is one of my most treasured belongings.

dropped from the belly of another plane. As I was to learn, these features reflected the challenges of entering an entirely new aerodynamic regime, as well as a bit of internal politics.

Before we get into all that, let's establish a few key concepts, starting with the speed of sound itself. We can think of sound as a disturbance, a form of energy produced by vibrating objects, and we're usually most interested in how that disturbance moves through air. As an object – be it a drum or a set of vocal cords – vibrates, it interacts with nearby air molecules, causing them to vibrate in response. Those molecules then jiggle the next-nearest air molecules, and the next-nearest, and so on. That ripple that moves through the surrounding air is what we know as a sound wave, and its speed depends on what it's travelling through. According to the basic equations of sound, two properties determine how fast it moves in a material: elasticity (stiffness, or resistance to deformation) and density (the number of particles in a given volume). But in a gas like air, it's better to talk about pressure, density and, most critically, temperature.

The pressure and density of a gas are proportional to each other – in other words, they increase and decrease at the same rate. A gas stored under high pressure also has a high density, because its molecules have been pushed closer together. But gas density strongly depends on temperature – the hotter a gas is, the quicker its molecules can whizz around and the further apart they can spread. This leads to a pretty surprising conclusion. A measurement of the speed of sound at the top of Mount Everest would produce the same result as a measurement taken at sea level … as long as the temperature is the same in both locations. The effect of the difference in air pressure that exists at these altitudes is cancelled out by the changes in air density. So when it comes to the speed of sound in air, what really matters is the temperature. The higher the air temperature, the faster sound can move through it, which means that if you really want to travel faster than sound, it's a good idea to seek out colder temperatures. On the day that Chuck Yeager wrote himself into the history books, he flew at an altitude of 13km (8 miles), where temperatures can reach -56.55°C. That high up, the 'local' speed of sound is close to 1,068km/h (664mph). Anything below that is considered subsonic, while anything above it is supersonic. Yeager reached a top speed of 1,127km/h (700mph). In high-speed aerodynamics parlance, that's equivalent to Mach 1.06, calculated by dividing the speed of the aircraft by the local speed of sound.

In *The Right Stuff*, the associated scene is tense and dramatic. As the Machmeter on the dial of the *Bell X-1* creeps ever closer to the elusive 1.0, the plane begins to buck and shudder. At 0.99, the pressure dial cracks, and the shuddering becomes violent. Then suddenly we hear a loud boom echo across the Mojave Desert. Yeager had done it. He'd gone supersonic. But we get a slightly different and more complete picture of the flight's aerodynamics from Yeager's autobiography, published in 1985. The buffeting, rather than happening very close to Mach 1, was worst at lower speeds, in and around Mach 0.88. In fact, the closer he got to his maximum speed, the smoother the flight was. Yeager admitted that in the moment the Machmeter tipped past 1.0, he felt both elated and numb, a little underwhelmed. 'After all the anxiety, breaking the sound barrier turned out to be a perfectly paved speedway … It took a damned instrument meter to tell me what I'd done. There should've been a bump in the road, something to let you know you had just punched a nice clean hole through that sonic barrier."* I heard similar sentiments from modern-day test pilot Wing Commander Marija (Maz) Jovanovich of the Royal Australian Air Force, when I asked about her first Mach >1 experience. 'When you're sitting in the aircraft, you don't even notice that you've gone supersonic. Everybody on the ground notices, but you don't. I was thrilled afterwards, of course, but honestly, the moment itself was a bit of an anti-climax.' So what's going on? How could the journey through this fabled barrier – one that still looms large in the public imagination – be so smooth?

Barrier

Part of the answer to this is that the sound barrier is less a hard, physical obstacle to flight, and more like a squishy and complex transition zone from one type of aerodynamic flow to another. As the earliest generation of aeronautical engineers discovered, critical changes to the way air flows around an aircraft start to happen long before its dial reads Mach 1. During the First World War, biplanes made largely from wood and fabric barely exceeded airspeeds of 143mph (230km/h), but their propellers, thanks to the combination of rapid rotation and forward motion, were much faster at cutting through the air; their tips sometimes reached supersonic speeds. That airflow could be so different on parts of the same object was curious, but it didn't really have any demonstrable

* All Yeager quotes are taken from the book *Yeager*, 1987 edition, pages 176–7.

impact on these 'slow' planes. A study carried out in 1918–20 changed all that.

By testing different wing sections, known as aerofoils, in a new high-speed wind tunnel, engineers showed that beyond about 350mph (563km/h), the lift forces acting on an aerofoil dramatically decreased, while the drag forces soared upwards. A sudden change like this would send an aircraft into a steep dive, one a pilot might not be able to correct. The same experiment showed that the thinner a wing was, the higher the speed it could reach before suffering the same fate. This and several studies that followed led engineers to conclude that air, which they had long (rightly) considered to be incompressible in their calculations, behaved very differently at high speeds. If an aircraft was fast enough, its motion through the air could compress the gas, actively changing its density. This might seem like a fairly niche conclusion, but the incompressibility of air – the idea that it would glide past an object rather than pile up on its edges – had been a key assumption of aerodynamics for decades. The realisation that it wasn't always valid is what opened the door to supersonic flight, because once it could be fully understood and measured, it could be designed around. By the early 1930s, planes that could reach 400mph (644km/h) were being built, thanks to improvements in engine design. Understanding – and counteracting – the effects of compressibility was no longer an academic curiosity. It was an urgent question that needed urgent answers.

The first thing to establish was *why* these changes were occurring. What was physically happening on the surface of the wing that would cause it to suddenly drop out of the sky? A group of engineers at Langley Research Centre (one of the earliest NASA facilities) started studying this breakdown of airflow in earnest. Schlieren photography – a novel imaging technique that can highlight tiny changes in the density of air as it flows around an object – revealed the secret. When they placed an aerofoil in a high-speed but still subsonic flow, a shockwave could be seen forming on its top surface, halfway back from the leading edge, and pointing almost directly upwards. In front of it, the boundary layer flow appeared to be laminar, but immediately behind it, the flow was turbulent. As they increased the airspeed, a shock formed on the bottom of the wing too, and both shocks got larger and moved rearwards. The region of separated, turbulent flow behind the wing grew ever larger.

The engineers determined that these shockwaves had caused the flow to catastrophically detach from the aerofoil. This explained not only the huge increase in drag, but also the reduction in lift. As air molecules met the aerofoil, they were accelerated across the top of its streamlined shape; at low speeds, this acceleration was the source of lift. But in this test, when the airspeed exceeded 0.6 Mach, some of these air molecules were pushed to supersonic speeds, and by colliding with other air molecules, they caused an abrupt change in the local air density – in other words, a shockwave. Its presence proved that a plane doesn't have to reach supersonic speeds to suffer from the effects of air's compressibility.

The name given to this condition, in which some localised areas of supersonic flow form on an object flying below Mach 1, is **transonic**. For a full aircraft, it can kick in anywhere from about Mach 0.75. Maz Jovanovich, speaking to me from an undisclosed location, told me:

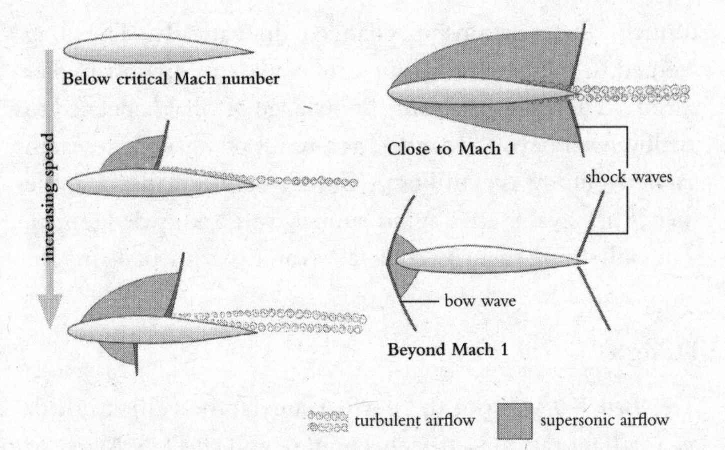

Figure 12: As the flight speed of an aircraft approaches the local speed of sound, shockwaves form and move along wing surfaces.

> When I think about the transonic regime, the word that characterises it for me is instability. This instability comes from a bunch of shockwaves – areas where the local flow goes supersonic – forming in weird places on the aircraft. It might be the engine intakes, the leading edge of the tailplane, the hinge line of the flight controls, etc. In an aircraft not designed for supersonic flight, shocks forming in those areas would cause a sudden loss of control.*

Transonic instability, and the wave drag it produces, is where the concept of a sound barrier really comes from. At low speeds, drag was known to increase with the square of the velocity (v^2), but at the speeds achieved in the Langley wind

---

* Maz Jovanovich really knows what she's talking about. She did her test-pilot training at Edwards Air Force Base, the very same place that Chuck Yeager tested the *Bell X-1*. Can you imagine how excited I was to talk to her?

tunnels, that relationship changed dramatically. The drag seemed to increase to infinity, 'like a barrier against higher speed'.* There was also some knowledge of what happened to airflow well beyond Mach 1, as a result of weapons research carried out by the military. But in between, beyond the speeds achievable in a wind tunnel, was a knowledge gap. The only thing that could fill it was an experimental aircraft.

Design

The *Bell X-1* was not the perfect supersonic vehicle. Built as a collaboration between Langley and the US Army, it was a lesson in compromise. For the Langley engineers, the goal of the *Bell X-1* research programme was to gather data to better understand the transonic regime. For the military engineers, the main goal was to fly supersonically – to demonstrate that the sound barrier could be overcome. In the end, the aircraft managed to achieve both. Filled with sensors and instrumentation, the *Bell X-1* was designed to give it the best chance of probing the unknown (nicknamed by Yeager and others as the 'ugh-known'). Its streamlined shape was based on a machine-gun bullet that had been shown to be stable at supersonic speeds, and it had relatively short, straight, thin wings. Its horizontal tailplane sat higher than its wings, to avoid any risk of it sitting in turbulent flow, and it could be deflected up and down to provide extra stability at high speeds. Its weight was kept as low as possible, and its propulsion system was

---

* British aerodynamicist W.F. Hilton said this in an interview with a newspaper journalist in 1935. The media story that followed represented what he said as 'the sound barrier', which may be where the phrase originates.

a high-powered rocket motor with multiple chambers. As Jovanovich explained, this powerful engine was especially important: 'The engineers realised early on that if they wanted to overcome this huge increase in drag, they'd need to get enough thrust, enough power, to punch through it.'

Over the course of 49 flights, this orange flying lab was methodically put through its paces. On flight #50, Yeager's plane entered the transonic regime, and as multiple shockwaves began to form on its wings and other edges, the drag increased, causing it to shudder through Mach 0.88. He adjusted the tailplane, and the rocket engine kept on pushing, driving the shocks towards the rear of the wing. Close to Mach 1, the flight began to smooth out. Engineers later learned that this was caused by the shocks joining at the wing's trailing edge – this decreased the flow separation, reducing the drag acting on the wings. The smoothness continued too, thanks to another shock called a bow wave that was forming, unknown to Yeager, just ahead of his aircraft's nose. At sufficiently high speeds, this bow shock suddenly attached itself to the sharp nose, forming a widening cone around the plane, stabilising the airflow around it. This marked the moment at which all of the flow around the *Bell X-1* became supersonic. On the ground, it was noted by a boom – likely the first ever heard by humans – as the cone of compressed air swept across the desert landscape.

With every flight of the *Bell X-1*, pilots and engineers learned more about the 'ugh-known', leading to continuous updates to the aircraft's design and that of its successors. Angling the wings so that they pointed back towards the tail, rather than sticking straight out from the fuselage, delayed the formation of local shockwaves, allowing the aircraft to safely reach higher speeds. Ensuring that changes to the cross-sectional area of the aircraft were gradual along its length helped to reduce wave drag. And while these started out as design decisions for experimental aircraft, their legacy lives on in today's commercial airlines, which, as Jovanovich explained, 'all fly in the transonic regime. The sweep in the wings, and the fuselage that follow the area rule are all measures taken to offset the increase in drag.' Transonic flight is now so well understood that 4.4 billion journeys were safely taken by passenger aircraft in 2018. Supersonic planes aren't solely experimental any more, with models produced by several major aircraft manufacturers, but they are currently very much the reserve of the military. Several private companies are aspiring to change that. Inspired by *Concorde*, which once reached Mach 2.04 at its cruising altitude above the Atlantic, they are aiming to bring back supersonic jets for wealthy business travellers. But whether the appetite for such a service outweighs the enormous financial and environmental cost of running it, remains to be seen.

As *The Right Stuff* showed, many more challenges lay beyond Mach 1 and 2. Overcoming them is what allowed humanity to reach beyond the Earth's atmosphere and return safely. But even with all our knowledge of thrust and drag, we couldn't have done that without first learning how to handle the heat.

Skin

It's not immediately obvious that extreme heating would be an issue for high-speed aircraft that fly at such lofty altitudes. After all, at 20km (12 miles) above sea level, air temperatures drop below -50°C. But on the surface of the aircraft, within the boundary layer, the situation is far from cold. We know that the air molecules stuck to the surface of any moving object are under constant bombardment from faster molecules further out into the flow; this is the source of skin friction. In low-viscosity fluids like air, such interactions can be incredibly turbulent and dynamic, and as airspeeds increase, the drag that results from them grows too. 'Skin friction scales with velocity squared, so the faster an aircraft flies, the more dominant it becomes,' said Professor Andrew Neely from the University of New South Wales. As billions of air molecules around a fast-flying aircraft bash into one another, they exchange kinetic energy, similar to the way a well-struck snooker ball causes others to scatter. Neely continued, 'That energy can't just disappear – it has to go somewhere. So, it manifests as heat.'

The link between friction and heating is one of the most influential partnerships in surface science, and it's something we'll explore more in later chapters. As aerospace engineer Dr Priyanka Dhopade told me, for high-speed aircraft it means that, 'You can't separate the airflow on the surface from the heat transfer side of things, any more. Rather than thinking about aerodynamics, you need to start thinking about aero*thermo*dynamics.' Material choice becomes all-important. At its cruising speed and altitude, *Concorde*'s skin temperature ranged from 91°C (196°F) in the rear section of the fuselage to 105°C (221°F) along the leading edges of its wings. Unsurprisingly, its nose was the hottest part of the aircraft, reaching 127°C (260°F) in flight. Finding a suitable set of materials that could retain their strength at these elevated temperatures while keeping mass to a minimum was a challenge. In the end, the majority of its airframe was made with a specially designed aluminium alloy that contained small quantities of copper, magnesium, iron, nickel, silicon and titanium. Faster aircraft, such as the

SR-71 Blackbird, which flew at Mach 3.2, used a mixture of titanium and polymer composite materials for its shell. But NASA's rocket-powered aircraft, the *X-15A-2*, was a whole different ball game. In 1967, it reached a top speed of Mach 6.7, equivalent to 7,274km/h (4,520mph). With it, aviation entered the hypersonic regime.

Neely, who is based in Canberra, told me:

> There is no exact point at which you switch from supersonic to hypersonic – there's no sudden appearance of shockwaves. Nominally, it happens at Mach 5, but the transition is fairly gradual. One of the things that marks the difference between the two is the extreme importance of heating. Skin friction is absolutely critical at hypersonic speeds, and it can account for half the total drag acting on an aircraft. Designing your vehicle to be robust enough to withstand the thermal structural heating that comes with that is a huge challenge.

The team behind the *X-15A-2* estimated that at its maximum velocity, parts of the aircraft could approach 650°C (over 1,200°F), hot enough to melt aluminium. A new heat-resistant nickel alloy called Inconel-X 750 was used to form the bulk of the airframe. It provided most of the thermal performance, as well as giving the aircraft a dramatic, gun-metal-coloured appearance. For its highest speed flights, two more coatings were added – one, a pink 'ablative' material that acted as a sacrificial layer of protection, gradually shedding off the surface in response to the aerodynamic friction. On top of that was a white sealant paint, which prevented the ablative from reacting with the liquid oxygen that powered the rocket engine. This combination of materials acted as a thermal protection system for the aircraft, and allowed pilot Pete Knight to fly at a record-breaking Mach 6.7 and land safely. But Knight's *X-15A-2* still bore many scars of high-speed frictional heating – the brakes and rudder were heavily charred, and even more seriously, part of a vertical stabiliser beneath the wing was entirely burned through.

The aircraft might never have flown again, but its development guided the burgeoning US space programme that was to follow. Among other things, it enabled the first direct measurements of hypersonic skin friction, and the discovery of localised 'hot spots'. Systems for flight control and stability, aerodynamic design and heat-shielding all came from the programme, as well as vital data on the effects of hypersonic flight on the human body. I think it's fair to say that the Apollo missions that carried astronauts to the Moon, the Space Shuttle that for 30 years ferried humans to Earth orbit, and the planetary landers that have given us a glimpse of other worlds, all owe a scientific debt to the *X-15A-2*. But even its top speed wasn't high enough to uncover all the mysteries of hypersonic flight. That honour goes to the blunt re-entry vehicles that slam through the atmosphere on their way down from orbit.

In space, drag isn't an issue. The lack of air means that spacecraft face no resistance on their journeys, allowing them to reach unimaginably fast speeds.* As you might imagine, that changes as soon as a spacecraft wants to land on a planet. The starting descent speed for the Space Shuttle as it entered the atmosphere was around Mach 25. For the Apollo capsule, it was in excess of Mach 30. So rather than try to minimise drag, re-entry vehicles need to maximise it, using its resistance to slow them down. The blunt shape drives up pressure

* At the time of writing, the Parker Solar Probe had reached a top speed of 244,226 miles per hour. That's 109.2 kilometres (68 miles) per second.

drag, and as Dhopade explained, it has another benefit. 'The bow shock that forms on a blunt body isn't attached to it like it would be on something sharp-nosed. It sits out in front, and acts as a sort of shield, an obstacle to the flow.'

Neely agreed, saying, 'The more normal or perpendicular a shockwave is to the surface, the bigger the temperature rise across it. So having a nice standoff shock keeps the hottest packet of gas away from the vehicle.'

Even so, the temperatures on a heat shield are still high, close to 1,650°C (> 3,000°F) in the case of the Space Shuttle. At those temperatures, air becomes a little less predictable. 'We can no longer treat air as an ideal gas, with a simple relationship between pressure, temperature and density,' said Dhopade. 'At hypersonic speeds, its chemical composition changes.'

Air's main ingredients – nitrogen and oxygen – are both diatomic, which means that they typically move around in pairs (N_2 and O_2). But at typical Shuttle re-entry temperatures, the bonds between these pairs of atoms break in a process called dissociation, leaving a cloud of highly reactive atoms close to the surface of the vehicle. For the higher entry speeds of the Apollo capsule, the chemistry is even more dramatic. 'You literally start knocking electrons off atoms, turning the gas around the vehicle into a sheath of partially ionised plasma that temporarily stops all communication with the vehicle,' said Neely. This gas-formerly-known-as-air pushes thermal protection systems to their limits. It means that they can't just be designed with extreme heating in mind; they also need to manage reactions that occur on their surfaces. Flying ultra-fast is as much about chemistry as physics.

Ablative materials remain the most popular option for spacecraft that need to pass through a planetary atmosphere. Applied in layers, they're designed to decompose in the presence of extreme temperatures. As they burn off, they dissipate heat, protecting the structure underneath in the process. The most recent Mars landers have all used a material called PICA (phenolic-impregnated carbon ablator) to survive the carbon-dioxide atmosphere, but other ablatives are under development.

★★★

For the most part, we can happily go about our daily lives not really worrying about the rules of aerodynamics. But I think that understanding them a little more can open a door, and give us a new appreciation for their implications. From sticking your hand out of the window of a moving car, to watching the precise curl of a match-winning ball, or marvelling over the fiery return of a vehicle from the depths of space, the interactions that occur between surfaces and the gases that surround them are ferociously complicated. But, as I hope you'll now agree, they can also be an utter delight.

CHAPTER FIVE

Hit the Road

The room was much larger than I expected, and dark, save for the huge, curved white screen that covered the far wall. But the thing that really dominated the space was sitting on a platform at the focal point of the screen: the partial chassis of a Formula 1® (F1) car. I was in Silverstone, visiting the factory of Aston Martin F1 Team (then known as Racing Point Force India). Officially, I was there to learn about the technologies that keep F1 cars stuck to the track, even when they're approaching the take-off speed of commercial aircraft.* But I was also lucky enough to get a sneak-peek at this, their high-tech simulator. 'The chassis is a few years old, but essentially the idea is to make the driver environment exactly the same as in the real car,' said Jonathan Marshall, my generous host for the day, and the team's Head of Vehicle Science. 'The driver puts their race seat in here. The steering wheel and everything else will work in exactly the same way as they do on track.' The platform itself, through means that I'm not permitted to describe, can simulate realistic motion of

---

* Depending on the weather and other factors, a fully loaded Boeing 737-800 reaches around 145–155 knots (~269–287km/h, 167–178mph) just as it lifts its nose to take off. In 2020, F1 champion Lewis Hamilton achieved an *average* lap speed of 264.362km/h (164mph) during the Italian Grand Prix weekend; at the time of writing, this was the fastest ever lap in F1. The *maximum* speed of an F1 car is considerably higher than this. For example, Hamilton exceeded 330km/h (205mph) at points on that same weekend.

the car, allowing the driver to feel some of the same g-forces they'd experience in a race. Having a set-up like this gives race engineers and drivers an opportunity to prepare before events, and to test their car even when they can't get access to a track. 'But what's arguably more important,' continued Marshall, 'is that before we've even made a new part for the car, we can "drive" it in here. We can get an idea if that change will really result in improved performance, which is ultimately our goal.'

This push for performance, and the search for almost imperceptible gains – tiny fractions of a second – is a hallmark of Formula 1, and whether you love the sport or hate it, there's no question that the engineering behind it is fascinating. Even a current season's F1 car is never really 'done'. Aspects of its design are tweaked and fine-tuned constantly, with teams using a wide variety of tools in the process. While simulation is among the most recent additions to their toolkit, it's certainly not rare – all of the big Formula 1 teams now have some sort of 'virtual car' set-up, and it's playing an increasingly important role. 'The motorsport industry spent a long time developing simulation tools that took variability – the driver, the environment and everything else – out of the equation,' said Marshall, as we headed down the corridor, leaving the simulator behind. 'In more recent years, that attitude has changed, largely because we now know that the system you're interested in isn't just the car. It's the car plus the driver on that track, in those conditions. This simulator allows us to change and control lots of things all at once, so it's really valuable.'

F1 cars have much more in common with the high-speed jets of the last chapter than any 'normal' road vehicle, because aerodynamics is truly all-important. Three key components allow a race car to manipulate the flow of air around it: the front wing, the rear wing and the diffuser. Both wings take inspiration from aircraft, using an aerofoil shape to generate lift. But because race engineers effectively mount these wings upside down, the 'lift' they generate is negative. So on an F1 car, the air passing under each wing travels much faster than the air flowing over the top of the wing. The result is **downforce**, a difference in pressure that pushes the wing down towards the ground. The angles of the front and rear wings are adjusted for each track, but together, they contribute somewhere between half and two-thirds of the car's overall downforce.*

The rest of the downforce comes from a diffuser, an open geometric structure on the underside (or floor) of the car. It's shaped so that as air flows into the space between the car floor and the track surface, it is accelerated through the narrow inlet of the diffuser, before spreading out in the widened, upturned outlet channels that sit at the rear of the car. The effect of this is that an area of very low pressure forms under the car, sucking it onto the road and providing downforce. With careful design, small, circulating pockets of low-pressure air called vortices can also be forced to form within the diffuser. According to legendary motorsport aerodynamicist Willem Toet, these

---

* It is difficult to find reliable figures for the relative importance of each component to a car's overall downforce, but that's not surprising. The car set-up changes track-to-track, each team uses a different car, and cars change every season. All of these have an impact on downforce. So I've gone with a general, very approximate value.

vortices don't just draw more air into the diffuser, they also 'mix it into the body of airflow that is expanding under the diffuser'. Just like with our dimpled golf ball, this mixing allows the flow to stay attached to the surface of the diffuser, improving the downforce even further. As I saw for myself in a different part of the factory, underfloor pressure is so vital to a car's performance that it is continuously monitored throughout each race. A spaghetti pile of thin plastic tubes connect an array of tiny holes in the expansive car floor to a sensor platform. This measures minute changes in air pressure, allowing engineers to spot any areas of unexpected – and potentially dangerous – flow separation.

Combined, these three systems – two wings and a diffuser – work to push the car down, sticking it onto the track with many times more force than it could get from its weight alone. This is why F1 cars can take corners at such blisteringly high speeds. Copse, one of the most famous corners in motorsport, is at the Silverstone Circuit, just a stone's throw from the Aston Martin F1 Team factory. Today's F1 cars zoom around that particular right-hander at speeds in excess of 290km/h (180mph).

But I'd be doing a disservice to you, my lovely readers, if I were to give downforce *all* the credit for the impressive cornering ability of these nearly-aircraft. Air alone cannot hold race cars onto the track, nor can it transform the power of their engines into forward motion. For that they need tyres.

Rubber

You're never more aware of the importance of grip than when you have none of it. A saturated bend, a patch of black ice, an engine oil spill; any of these can be enough to make a driver or rider lose control. Because tyres are the only contact point between a vehicle and the road, they govern much of that vehicle's behaviour. The effectiveness of braking, acceleration and steering are all influenced by what happens down where the rubber meets the road. But to define exactly how 'grippy' a tyre is, we need to consider a few different factors: the materials and design of the tyre, the condition of the surface it's travelling on, and the quality of the contact between them.

Let's start with the tyre.

'Managing the performance of race tyres is a bit of a black art,' former F1 tyre engineer Gemma Hatton told me. 'And one reason for that is the nature of tyre rubber itself.' Modern tyre rubber is usually a blend of natural polymers extracted from rubber trees, and synthetic polymers made largely from fossil fuels (tyres are *not good* for the environment).* Once it's been fully processed, this combination produces a flexible material that behaves somewhere between an elastic solid and a very viscous fluid. This property of rubber – its viscoelasticity – is both its superpower and what makes it a little unpredictable.

In an elastic solid like a spring, there's a simple, linear relationship between the size of a force applied to it and the response of the solid: the harder you compress a spring, the

* This is starting to change, very slowly, with more eco-friendly versions of synthetic rubber under development. Chemists at the University of Minnesota have developed a way to produce isoprene – a key ingredient – from glucose rather than petroleum. And tyre manufacturer Continental now makes some of its tyres with a rubber, dubbed Taraxagum, extracted from the roots of a fast-growing dandelion plant.

shorter it will get. You also know instinctively that when you release a spring, it will immediately bounce back to its original shape and size. A viscous fluid is very different because the friction that operates between its molecules works to resist any deformation you try to apply to it. Think about using a hand pump to inflate a ball. When you push down on the handle sharply, you'll find that the piston won't move instantly. Equally, when you release the handle, it won't spring back into its starting position. There is a lag, a time delay, between applying a force and seeing the resulting deformation. In materials science, this lag is called **hysteresis**, and it happens because some of the energy you used to apply the force (for example, to push the handle down) has been lost to the fluid in the form of heat.

A viscoelastic material such as tyre rubber straddles those two behaviours. As long as it's operating within its limits, rubber will deform and return to its original shape (like a spring), albeit with some time delay and a bit of energy loss (like a viscous fluid). How spring-like or viscous a specific rubber compound is depends not only on its composition, but also on the temperature at which it operates, and the types and speed of deformation it experiences. This is why there isn't one single tyre compound – in fact, it's very typical for a tyre to contain multiple different formulations within it.

In addition, filler materials such as silica and soot-like carbon black are added to tyre rubber to give it some wear resistance, as well as its distinctive colour. Plasticisers usually feature too – they make the rubber softer and more flexible, so the amount added will vary depending on the fate of the compound and the tyre it will eventually go into. By playing with the chemistry, designers can produce specific compounds that reach their peak of performance in the same conditions they'll face in the real world. At least, that's the idea. But as Hatton hinted, tyre formulation – especially the sort required for the upper echelons of motorsport – is more art than science. Every track and car is unique. Every driver has their own style. Every day brings slightly different weather. And F1 tyres also need to wear down predictably during a race, so that teams can have a strategy, including when to take their pit stops. Designing compounds to suit all of these variables perfectly is next to impossible. At the time of writing, Formula 1 used only one tyre provider, Pirelli. It's fair to say that they have their work cut out for them.

On top of that, blended rubber is just one of hundreds of ingredients that make a race tyre. Despite being rather different beasts in practice, motorsport tyres and road tyres are assembled in a fairly similar way. It starts with applying a sheet of an airtight synthetic rubber to a rotating drum. On top of this, two layers of 'textile ply' are added. Made from carefully aligned and rubberised nylon and polyester fibres, these plies are oriented so that the fibres overlap and form a dense mesh that reinforces the tyre. Strips of a flexible rubber are added to the edges of the rubber sheet – these are the sidewalls – along with another textile layer and a pair of high-strength metal hoops known as the bead. This rubber-ply sandwich is inflated on the drum to form a proto-tyre. On another machine, two belts made from more than a kilometre of fine steel wires and (you guessed it) rubber are overlaid so that the metal forms another strengthening criss-cross structure. The steel belts are added to the inflated carcass, and the whole thing is wrapped in nylon. A final layer of rubber, called the tread, is rolled onto the outer surface, leaving you with something that looks very like a tyre, but which is still soft and slightly misshapen.

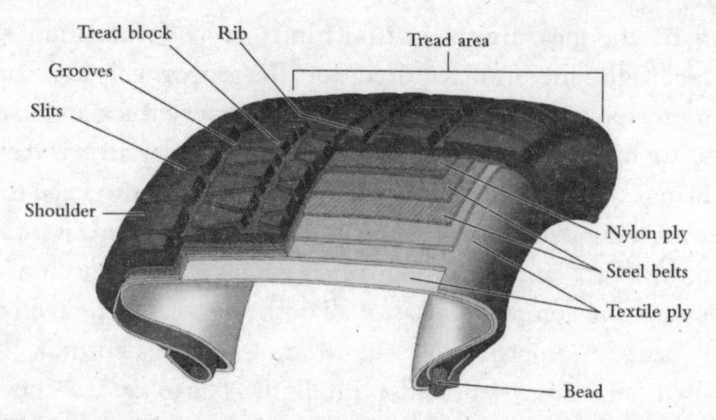

Figure 13: This cross-section of a road tyre shows that they're far more than just rubber rings.

The final step in a tyre's manufacture is vulcanisation, and this is when its properties get locked in. The tyre is effectively cooked at high pressures and temperatures in the presence of sulphur gas, which causes new links to form between the rubber compound's wobbly polymer chains. This increases the viscosity of rubber, making the tyre harder and more rigid. If this curing process happens in a smooth-walled metal mould, you'll be left with a very smooth tyre – a racing slick, most often used in F1. If it happens in a precisely engraved metal mould, its patterns are transferred into the rubber, creating a treaded tyre, like those used on road vehicles.*

For every form of driving, there's a wide range of formulations and tread patterns in use, with each combination suited to a specific set of demands. Even across motorsport, Hatton said, 'The differences [in tyres] are absolutely huge. F1 tyres really only have to worry about

* F1 cars can't always race on slicks. Sometimes they need 'intermediates' or 'full wets', which have tread patterns on their tyre surface that help to displace the water and keep the rubber on the road. More on this shortly.

tarmac – and the tarmac is most often dry. In WRC [World Rally Championship] which happens on gravel, dirt, snow and ice, the surface changes with every wheel that goes over it.' Regardless of the demands of the event, a tyre's main job stays the same – to grip onto the road as efficiently as possible, and give the driver the confidence to know that they'll make the corner.

Grip

At the heart of all this is friction. To understand where it comes from, we first need to zoom in on the interface between a treaded tyre and the road, so that we're looking at a small block of rubber in contact with a road surface. This portion of a tyre is called the contact patch, and it is critical in all questions about tyre performance.

The first thing you'll notice is that the road surface isn't smooth. Rather, it's heavily textured, covered in tiny bumps and dips. Because of the way roads are built – a mixture of crushed rock and bitumen – even a well-maintained racetrack will feature bumps that are somewhere between 1 micrometre and tens of millimetres in size. This surface roughness, combined with the viscoelastic properties of rubber, provides the first grip mechanism that tyres rely on. Each time the rubber block meets a rough spot, it deforms and 'flows' over it. But because of hysteresis, the rubber doesn't instantly bounce back after the deformation; it lingers, generating a resistive, frictional force between it and the road. This process, whereby the roughness of the road physically penetrates into the tread rubber, is called **indentation**, and it can provide a tyre with some grip even on wet roads.

The same is not true for the second friction mechanism that's at play. **Molecular adhesion** requires the rubber to

Figure 14: Even though moving tyres rotate, the flattened contact patch is always stationary with the respect to the road surface. The behaviour of the tyre in that patch controls much of its behaviour.

make intimate contact with the surface. According to tyre-maker Michelin, the distance between them should be no more than 1 nanometre, and that's because tyres want to exploit van der Waals forces – the very same forces geckos use to scale walls and ceilings (see Chapter 2). As rubber moves across the surface, vdW bonds are continuously formed, stretched and broken. This cycle, again made possible by rubber's viscoelasticity, generates a frictional force between the rubber and the surface, allowing it to grip.

The presence of water, however, interrupts the formation of these bonds, inhibiting the helpful friction they provide. So in heavy rain, tyres can't access molecular adhesion. But on a smooth, dry road surface, it's very important. Sometimes, during a motorsport race, you'll hear drivers talk about 'laying down some rubber'. What they're really doing is manipulating molecular adhesion. The contact between their tyre rubber and the track is so intimate that, as the bonds form and break, tiny pieces of the rubber are torn away from the tyre and remain stuck to the track.*

* This thin layer of rubber left behind by a slowly degrading tyre can help to generate even more grip for the tyres with each passing lap. In addition to rubber-on-road, it provides a rubber-on-rubber interface.

Although at a glance, these two mechanisms seem very different from one another, they share some important similarities. The first is that they both rely on the micro-movement, or slippage, of rubber over the surface. Without those rapid, imperceptible slips within the contact patch – be it the flow of material over a bump, or the stretching and breaking of bonds – rubber can't generate the friction it needs to stick to the road. Both mechanisms are also strongly temperature-dependent. Higher temperatures make rubber stickier and more flexible, and up to a point, this will maximise the grip that can be achieved via molecular adhesion *and* indentation.

If you've ever watched an F1 race, the obsession with tyre temperature will be familiar to you. Each rubber compound made by Pirelli has its own optimal operating temperature, which is always much higher than it would be for road cars. So while a race car sits on the starting grid or in the garage, all four of its tyres are shrouded in heated blankets set to temperatures in excess of 80°C.* When a driver comes in for a pit stop, any new tyres they get will also have been kept warm. 'To maximise the amount of grip, F1 tyres are soft,' said Hatton. 'This means they heat up quicker and therefore become more "sticky" to the track.' Once the car is racing, its tyres stay hot due to the friction between them and the track. This is where race tyres are at their happiest. But when those temperatures drop below or exceed their ideal temperature

* The storage conditions for race tyres – temperatures and durations – are tightly defined by FIA (the sport's governing body) and Pirelli. At the time of writing, blankets for rear slick tyres could be set to 80°C, and those for front slick tyres to 100°C. There are plans to phase out tyre blankets eventually, which if/when it happens, will have a significant impact on car design.

range, things get tricky. In the case of an incident on the track, all cars must follow behind the safety car and slow down. Every moment of driving at lower speeds cools the tyres, reducing their grip. Drivers try to maintain temperatures by weaving and swerving, but this only helps a little. As a result, as soon as the track has been cleared and they're free to race again, drivers are left with temporarily compromised tyres – cooler, stiffer and considerably less grippy. This makes their cars slower and harder to control, which can lead to a rather tense restart. It is also possible to overheat an F1 tyre, so that its rubber becomes extra soft and sticky. When track temperatures are higher than expected, the gradual, relatively stable degradation that gives us pit stops and exciting racing is replaced by a turbocharged version. Tyres rapidly deteriorate and friction drops away, as they leave more and more rubber behind them on the track. There's no way back from the grip 'cliff' – the only solution is a new set of tyres.

It's not surprising, then, that race teams and tyre engineers want to continuously monitor tyre temperature throughout a race weekend. But as Hatton explained, the process is not straightforward. 'What you really want to measure is the tyre bulk – inside the rubber – because this is the most accurate representation of the tyre temperature. However, this is impossible.' Teams have to compromise, and measure what they can. She continued, 'Each rim is fitted with a tyre-pressure monitoring sensor, which not only monitors the pressure, but also has an infra-red sensor that points to the underside of the tyre carcass.' Engineers can also measure the outer surface temperature of tyres, via infra-red sensors mounted on either the underfloor or the front wings. This, Hatton cautioned, is only indicative:

'Surface temperature fluctuates massively, so it is not the most reliable measurement. Carcass temperature is often what the engineers focus on, though every team has their own approach. The various measurements they get from their sensors help them to define the tyre's "optimum working window".'

During a race, tyre engineers from Pirelli and the teams remotely monitor the carcass and surface temperatures for all four tyres in real time, thanks to the wireless transmitters connected to each sensor. They get some extra hands-on information during pit stops, when tyres are taken off the car. 'Once a set of tyres has been used, the Pirelli engineer works with the tyre fitter to scrape off any debris. You can then use a probe to measure the depth of several tread holes that span the width of the tyre, and that gives you the wear profile.' This allows the team to identify any areas of uneven wear or degradation, and helps the Pirelli engineers to predict how the new set of tyres is likely to perform for the remainder of the race. 'The predicted number of laps is a useful guide for the team strategist. But the bigger teams will also be running hugely complex live tyre models.'

Even armed with this information, race tyres are still full of surprises. A brief foray off-track can see tyres pick up debris that stifles their stickiness. Hitting a kerb causes an extreme increase in tyre pressure, and can sometimes scratch or cut the tread. Excessive or sudden braking into a corner can cause the tyres to lock up, producing a flat spot on the tread that grips very differently from the rest of the tyre. F1 racing pushes tyres – and our understanding of them – to their absolute limits. Unless you're doing something you shouldn't, the tyres on your bike or car won't experience anywhere near that level of punishment.

Fundamentally, though, all tyres are governed by the same behaviours and interactions. All that changes between tyre types is which interaction takes top priority.

Tread

Picture the scene. You're sitting in your dream (ideally, electric*) car. The road ahead of you is smooth, and it gently winds its way through the landscape. You have nowhere to be, so you can take your time. As you drive, you're aware of the contact your tyres make with the asphalt. Their performance can be described by three key properties: grip, wear and rolling resistance. Until now, we've mainly talked about the first two. The third, which is sometimes called rolling drag, is primarily caused by the deformation that tyres experience under the weight of a car. We know that rubber is flexible and viscoelastic; it's what makes it so good at clinging to the road. But it also means that as a tyre rotates under load, the rubber in its tread and sidewalls undergoes repeated cycles of large-scale deformation and recovery, each time dissipating some of that energy as heat. As a result of this loss of energy, a free-rolling tyre will gradually slow down and eventually stop. To keep a tyre moving you need a

---

* I have a very complicated – possibly contradictory – relationship with cars. As you can probably tell, I love the engineering and technology of motorsport, and I regularly watch F1, V8 Supercars and, increasingly, Formula-E. However, I'm no great fan of private car ownership, nor of the continued priority given to fossil fuel-belching vehicles in the majority of our urban areas. I resisted owning a car until I moved to New Zealand, aged 33. Even now, driving is a last resort for me. If I can walk or take public transport to my destination, I do.

continuous input of energy, and in a car, that means fuel. The higher the tyre's rolling resistance, the more fuel you'll use, and while this isn't really an issue for race cars, it certainly is for passenger vehicles. According to the US Department of Energy, a 'gasoline [petrol] car' uses 4–7 per cent of its fuel just in overcoming rolling resistance, with much higher rates for trucks and other heavy vehicles.

Manufacturers use some tricks to reduce rolling drag, without adversely affecting the grip or wear of a road tyre. Adding silica to rubber compounds makes them slightly harder and less susceptible to the rates of deformation they'll experience on the road. In addition, the metal and textile belts within the carcass provide some structural rigidity. Cyclists among you will also know of a way to keep rolling resistance low – make sure your tyres are fully inflated to an appropriate pressure. This minimises any floppiness in the rubber, reducing the energy needed to keep the tyre rolling.

One design feature that plays a key role in the life of a road tyre is arguably the most obvious one: its tread pattern, made from a combination of raised ribs, angled blocks, deep grooves and narrow slits. The difference between those tyres and the pattern-free slicks favoured by F1 are easy to see. But while the latter offer masses of grip and remarkable handling performance thanks to the huge patch of rubber that makes contact with the road, they're pretty useless on anything other than a hard, smooth track. Slicks also wear down incredibly quickly, rarely lasting beyond half a race. Road tyres simply can't work that way – they need to provide reliable grip, adapt to changing conditions, and be robust enough to last for years of daily use.

To understand how tread patterns help road tyres to do all that, let's look at the tyres of your dream car as you apply the brakes while driving along the perfect road. The first thing that happens is the rolling speed of the tyres begins to decrease, and the weight of the car shifts forward, putting more load on the front tyres than on the rear. Zooming in further to where the front tyres touch the road, this then bends and deforms the flexible rubber blocks that make up the tread, especially those blocks that are towards the front of the contact patch. As the rolling continues, the tread blocks towards the back of the contact patch begin to slip, activating the tyre's two grip mechanisms (indentation and molecular adhesion), and slowing the rotation further. The more movement that can happen between each individual tread block, the more effective the braking.

What marks the transition between tread deformation and micro-slippage is the coefficient of friction (μ) of rubber-on-road. As you hopefully remember from way back in Chapter 1, this coefficient is a measure of how much force is required to make one material slide relative to the other material. For a rubber tyre on a dry road, μ is always somewhere between 0.9 and 1.3, which means that there is a very good contact between them. In wet conditions, μ drops considerably, and can reach values as low as 0.1 if the road surface is super-smooth. Thankfully, most roads in the real world are rough enough to retain some of their grip properties in the wet. But in order to access that texture, tread rubber needs to be able to flex, as well as to push through the water. Luckily, your dream car is ready for a sudden downpour – its tyres are in perfect condition.

None of the patterns on a tyre are random. They all have a specific job to do. As a tyre rolls over a wet surface, a small bank of water can build up in front of it. Slits in the tread splay out, suck this water up off the ground and direct it into wide grooves that are cut around the circumference of the tyre. From there, the water is channelled into lateral grooves that force it out the sides of the tyre and away from the contact patch. This allows the tyre's raised ribs and tread blocks to make direct contact with the road, tapping into indentation and providing grip. Tyres are amazingly efficient at removing the water from a wet road. Pirelli says that the heavily patterned wet tyres it produced for the 2020 F1 season 'can each disperse up to 65 litres of water per second' when the car is travelling at full speed.

Tread patterns are similarly useful each time you drive around a corner. As you turn your steering wheel, your tyres point towards the bend, and just like with braking, the weight of the car will then be unevenly distributed; this time the load is higher on the outer two wheels than on the inner two. In cornering, the flexing and micro-slipping of the tread blocks that give a tyre its grip happen laterally – in other words, tyres deform sideways. So a pattern that is equally grippy in all directions, combined with a rigid tyre sidewall, will give you the most reliable cornering behaviour.

The only real downside to tread patterns is how loud they can be. Believe it or not, most of the noise you hear from cars driving at highway speeds is produced by the interaction between the tyre and the road. Those same interactions are also a significant noise source for heavy vehicles driving at or above 50km/h (30mph). Deformation of the tyre, and the flexing, release and micro-slippage of the tread blocks on the textured road surface all generate sound, and can be influenced by the weight and speed of the vehicle. The tyre itself, with its flexible cavity filled with air, can act like a drum as it rolls along, producing a low-frequency hum. The size and orientation of the tread blocks, and the gaps between them, are also likely to have an impact on tyre noise. Generally, the chunkier and more textured the tread pattern is, the noisier it'll be, although racing slicks are far from silent. The road, too, plays its part, with smooth, porous surfaces producing the lowest air pressure under and around the tyre, resulting in a reduced noise level.

As with most things in this book, it's impossible to separate out all these effects and interactions from one another, but I'd argue that that's what makes tyre grip so fascinating. Rubber underpins it all. The material, used by humans for more than three millennia, utterly revolutionised wheeled transport when the two first came together in the 1800s. Tyres gave us a way to move things smoothly and efficiently; to use friction to our benefit and propel us forward at speed. But as we'll soon discover, friction also has the ability to make us grind to a halt.

Brake

Bertha woke before dawn. She had a long drive ahead, so wanted to set off early. She and her two teenage sons, Richard and Eugen, quietly made their way to the garage to get the car ready. The route, from their home in Mannheim to Bertha's birthplace of Pforzheim, would take them south, with lots of stops along the way. Thankfully, the weather was fine on this particular August day, because their car had no roof. Its wheels, too, were fairly rudimentary; the two rears were lined in steel, while the single steerable front wheel was lined in solid rubber.[*] But despite its simple appearance, this car was an engineering marvel. Designed and built by Bertha's husband Carl – with her financial backing – the *Benz Patent-Motorwagen* was the world's first production automobile.[†] Carl Benz was a ferociously talented engineer. In just a decade, he patented designs for the petrol two-stroke engine, its speed regulator, ignition system, carburettor, and an evaporation-based cooler, as well as his *Motorwagen*. But he had no head for business, and preferred tinkering in his workshop to drumming up publicity for his new invention.

Bertha, on the other hand, knew that in order for their company to be a commercial success, people needed to hear all about it. She hatched her plan, and on that day in 1888, she and her sons headed off on their journey – the first long-distance automobile drive in history. They left a note for Carl which said that they were going to visit Bertha's mother in Pforzheim, but omitted to mention the means by which they were travelling. It would be several hours before Carl realised what had happened.

It's hard to overstate the scale of Bertha's undertaking. Until then, the vehicle had travelled no more than a few hundred metres on paved paths, and always within reach of its designer. It stored just 4.5 litres of fuel (about 1 gallon) and relied heavily on regular water top-ups, which

---

[*] The *Benz Patent-Motorwagen Model I* had steel-spoked wheels and solid rubber tyres. Bertha Benz drove the *Model III* – according to Daimler (the parent company of Mercedes-Benz), its wheels had wooden spokes and were lined in either steel or rubber.

[†] Benz's name is spelled Karl and Carl alternately. I went with the spelling that appeared on the patent for his Motorwagen.

would mean the trio would have to find supplies en route, in a time long before service stations. And it had just two gears; not quite enough to make it up the steep hills they'd meet along the way. Despite all this, less than 12 hours after they'd set off, Bertha sent Carl a telegram to confirm that they'd arrived safely, having travelled 104km (65 miles), with only a limited amount of uphill-car-pushing.

As a piece of publicity, the journey was remarkably successful. The press were fascinated by Bertha, her boys and the 'smoking monster' built by her husband that bore the trio safely to Pforzheim. Orders for the *Motorwagen* began to flood in. The Benz family was on its way to fame and fortune. But the journey also played a critical role in the technical development of the car. It was the first true 'test drive', and Bertha solved multiple practical problems along the way, like repairing a damaged ignition wire and unblocking a fuel line, with only the things she had on her person. The drive encouraged Carl to add another gear, which helped to make the car infinitely more practical. Bertha also invented something absolutely central to road vehicles: brake pads. The brakes in Carl's original design were made from solid wood, but Bertha realised that they would be much more effective at gripping onto the steel wheel rims if the wood had a flexible but robust outer coating. So for their return drive to Mannheim a few days later, she asked a local cobbler to clad the pads in thick leather. The resulting brake pads became the standard offering for the *Benz Patent-Motorwagen*, capable of slowing down the vehicle from its top speed of 16km/h (10mph).

But as engines became more powerful, cars got faster, heavier and harder to stop. Leather-on-wood brake pads made way for rubber-impregnated cloths and fibres, which came with their own issues, not least a tendency to burst into flames just when they were most needed. The first decade of the twentieth century offered a solution – a material that was non-flammable, chemically resistant, strong and relatively cheap. Asbestos, despite its link to significant health hazards, remained the brake pad compound of choice for more than 70 years.* Today's brake pads are made from a wide variety of materials, which can largely be categorised into binders, fillers and friction modifiers. The binder is the glue that holds everything together, and it's most often made from phenolic resin. You can blame this for the acrid smell produced by overheated brakes. The fillers, which could be anything from metal fibres to ground-up rubber, are chosen to suit the specific needs of the brake pad: perhaps increasing strength and durability, or reducing the noise produced by the brakes. As their name suggests, friction modifiers fine-tune the frictional properties of the final brake pad. Powdered metals are widely used for this, as are ceramic powders, blended with graphite and nut extracts, with each playing a specific role in the mix. Ultimately, though, brake pads are designed to do one thing – to generate consistent, reliable friction.

In the *Motorwagen*, the driver had to pull a lever to engage the brakes, pressing the pads onto the rear wheels. Modern car braking systems are significantly more sophisticated,

* Warnings around the harmful effect of asbestos on human health have been around since at least 1898, and since then its fibres have been linked to cancers and other dangerous lung conditions. At the time of writing (2020), all forms of asbestos have been banned in 67 countries worldwide.

but the outcome is largely the same – contact between the brakes and the wheel makes the car slow down. In a disc brake, a pair of brake pads is mounted straddling a flat disc (or rotor) that rotates with the wheel. These pads normally sit away from the rotating disc, but when you put your foot on the brake pedal, small hydraulic pistons push them against the face of the disc. Drum brakes work the same way, though the pads and the surface they're pressed onto are curved. Both systems will bring a vehicle to a stop, but brake discs do it more efficiently. That's why they're the default option for most road cars – at least for the front two wheels, which do the bulk of the braking. Drum brakes are usually confined to the rear wheels, but in an increasing number of high-performance cars, they've been replaced by discs there too.

So what actually happens when a brake pad comes into contact with a disc? The first and most obvious thing is that a lot of heat is produced. Brake temperatures beyond 1,000°C are fairly typical in an F1 car, which is why you'll sometimes see the discs glow as a driver decelerates into a corner. The usual explanation for where this heat comes from is that it's generated by the friction between the pad and the disc, but it's a *bit* more subtle than that. A moving car has a lot of kinetic energy, so to slow down, it needs to shed that energy, eventually reducing it to zero. Disc brakes do this by converting the kinetic energy into other forms – mainly heat, but also sound, and occasionally light. As the brakes are applied and the pad and disc begin to slide along one another at speed, their surfaces interact. The brake pad, thanks to its long ingredients list, is designed to be rough and textured. But even the polished brake disc, which in a standard road car is made from cast iron, isn't perfectly smooth. So their microscopic surface features bash into one another, sometimes bending and flexing, other times cracking and breaking. This not only produces a resistive (frictional) force that works to slow down the disc, it also dissipates that kinetic energy.

So in a way, high brake temperatures mean the process is working, but if left unchecked, they can cause lots of problems. When brake pads are manufactured, they're 'cooked' at high temperatures and pressures that cure the binder and bond the friction material to the backplate. If the brakes are applied repeatedly or for a sustained period of time, they can get so hot that the binder starts to evaporate. The escaping gas forms a thin layer between the pad and brake, reducing their contact and causing a drop in friction. This is one form of brake fade, and it can be a pretty scary experience – despite pressing firmly down on the brake pedal, you get only a small reduction in speed. Excessively high temperatures can also cause the hydraulic fluid in the brake lines to boil, which again makes the brake pedal less responsive. Brake fade is usually temporary, and can be rectified by lifting off the pedal to allow the brakes to cool down. Disc brakes are less susceptible to fade than drum brakes, for the simple reason that they're open to the air, so they can cool down fairly quickly. Material choice can also have an impact – compounds extracted from cashew nut shells increase the thermal stability of brake pads, helping them to dissipate generated heat. In addition, some road cars and heavy vehicles use grooved or vented discs. They give the binder gas an escape route, and increase the airflow around the braking system. In F1, where the speeds and temperatures are high, and every component is pushed to extremes of performance, brakes get a major upgrade.

Race

It might seem counter-intuitive, but braking is a crucial part of driving fast. On any racetrack, one of a driver's goals is to stay on the racing line – the shortest path around the track. So rather than follow the long outer curve of a tight bend, as they turn they'll instead 'clip' the inside of the curve, a point known as the apex, to minimise the distance they have to travel.* Doing this requires very precise braking: putting *just* the right amount of pressure on the pedal for *just* the right amount of time. When they manage it, a driver will emerge from the corner in a great position on the track, still carrying the speed they need to tackle the next part of the course. But driving like this takes its toll on the brakes. And on some tracks, brakes don't get much of an opportunity to cool down.

Take the world-famous Monaco street circuit. At just 3.34km (just over 2 miles) long, it's the shortest track in the F1 calendar, but it involves lots of braking and accelerating. According to brake manufacturer Brembo, drivers in the 2019 event used their brakes for 18.5 seconds per lap, which is more than a quarter of the total lap duration. At the most demanding corner, cars decelerate from 297km/h (185mph) to 89km/h (55mph) in less than 2.5 seconds. That's a lot of kinetic energy to rapidly transform into heat. It's no wonder brake discs glow. To manage this enormous heat load, manufacturers drill tiny radial holes – more than a thousand of them – into the edge of each disc. The holes increase the surface area of

---

* This isn't an option for daily life. Lanes, rather than the need for speed, are typically what define a vehicle's position on the open road.

the disc, making it easier for heat to radiate away. But they also act as ventilation. When combined with large cooling ducts mounted onto each wheel rim, they draw cool air into the centre of discs, and carry hot air out at the edges. And as a bonus, these F1 discs are incredibly light, weighing in at 1kg (2.2lb) each, compared to 15kg (33lb) for a similarly sized cast-iron disc. So why aren't we all using them? One reason is price – each disc can cost upwards of $2,000 (about £1,500) and take six months to make. They don't last very long either, and are generally replaced after each race. And finally, they operate within a very specific temperature window, 350–1,000°C. Below that minimum, they offer almost no stopping power – the pad and discs just can't produce enough grip. But if they experience temperatures above the maximum for too long, they can fail catastrophically. As Jon Marshall described it to me, 'it's as if you start machining through the material. It's unbelievable, how quickly a brake disc can run out of "meat" when that happens.'

Technology can help a team and driver to manage their brakes, but as with most things in F1, it's not straightforward. The size and shape of the cooling ducts control how much air gets forced through the discs, so you'd imagine that bigger would be better. But as legendary F1 engineer Pat Symonds told *Racecar Engineering* magazine, cooling has consequences: 'for a heavy braking circuit such as Montreal, we are forced to use some of the biggest ducts of the season. Moving from the smallest to the largest cooling ducts can cost 1.5 % in aerodynamic efficiency, which represents a loss of 1km/h in top speed.' I'd imagine this has led to a few arguments between a team's brake engineer and their aerodynamicist. Even measuring the temperature of the brake assembly is challenging. Marshall told me that at

Aston Martin F1, they use a combination of high-temperature thermocouples embedded in the brake pad's mounting bracket and a series of infra-red sensors pointed directly at the disc. The colourful thermal images that occasionally feature on TV race coverage are mostly there for us viewers – the temperatures they suggest are indicative at best.

The other important process that happens between a brake pad and a disc is wear. All that sliding and rubbing causes physical damage to both surfaces; particles are torn from them each time the brake is engaged. Over the lifetime of a braking system, this gradually reduces the coefficient of friction of the materials – in other words, they lose their grip. But it's not just because the surfaces are 'polishing' one another or are losing their thickness. Wear also generates something called a tribofilm – a very thin layer of fine-grained material that is crushed up in the contact between the pad and disc. 'Tribofilms are very influential when it comes to wear and friction,' said Dr Shahriar Kosarieh from the University of Leeds. 'We consider the film a "third body", because despite it being made from the two materials that are sliding against one another, its chemical and mechanical properties are different from either one.' German researchers who looked at a variety of commercial brake pads on cast-iron discs found that, regardless of what the pad was made from, the tribofilm that formed was always dominated by iron oxide (Fe_3O_4), with the other ingredients playing a fairly minor role. 'The tribofilm controls the dissipation of heat and can reduce friction – it dominates the performance,' Kosarieh continued. 'Brake manufacturers know this, and consider it when they're designing their pad formulations. Pads and discs are matched together to give the best performance.

And if you change either material, you'll change what's produced at the interface.'

In Kosarieh's recent research, he looked at the frictional behaviour of lightweight alternatives to cast-iron brake discs, mainly those based on aluminium. He's not the only one – the entire automotive sector is obsessed with reducing weight, mainly because the lighter a vehicle is, the less fuel it will consume, and the lower its environmental impact. Aluminium is currently leading the charge. 'It is a low-density metal, about 2.5 times less than that of grey cast iron, so the potential for weight reduction is huge,' he said, chatting to me on the phone. 'It has high thermal conductivity too, and the oxide that forms on its surface can offer corrosion protection.' By combining aluminium alloys with a hard ceramic material like silicon carbide, you can add high strength to that list of properties too. 'But the problem with aluminium is that at temperatures above 400°C, it could start to melt. In brakes, that would mean a sudden drop in friction, which is the worst thing you could have. So there's a big push to find ways of engineering the surface to give it more thermal stability and make it last longer.'

For Kosarieh, one of the most interesting approaches is plasma electrolytic oxidation (PEO) which uses an electric field to grow a complex and highly wear-resistant layer on the surface of aluminium. When he tested the performance of various PEO-treated aluminium discs, he found that some survived to temperatures of about 550°C. However, in a number of cases, the coefficient of friction was too low – below the required minimum threshold for a practical braking system. Kosarieh was undaunted. 'Brakes work as a system. If you have a new disc, then you need to optimise the counterpoint as well. Companies are designing new brake pad formulations specifically to work on PEO-coated discs.' I could only find a couple of published studies that combined PEO discs with these new friction pads, but the results look fairly promising. Lightweight aluminium brakes may well make an appearance on future road vehicles.

F1 found a different solution for their discs and pads in the late 1970s, and they've stuck with it ever since: a material called carbon-carbon, highly ordered carbon fibres embedded in a graphite matrix. And it's so good at dissipating heat, it was used on the Space Shuttle. While it might sound similar to the carbon fibre that forms an F1 car's chassis, it's actually a very different beast. Manufacturing carbon-carbon is slow and complex, with the material built up layer by atomically thin layer. Friction-wise, it's a winner, providing twice as much grip (within its optimal temperature range) as conventional brake assemblies.* But it's not magic. Under the stresses of racing, it will eventually wear down, partly via tribological means, but also chemically. At elevated temperatures, carbon-carbon reacts with oxygen in the air, which supercharges its degradation. It's the source of the black dust that you sometimes see when an F1 driver hits the brakes hard.

This process means that teams need to monitor their brakes for more than just temperature. Marshall told me that they use pressure sensors to keep an eye on the airflow through the ducts. For wear, they have electrical sensors that can measure lateral movement. 'We use these to measure how far the pad has to travel to contact the disc. From that we can infer how much total wear we have – that's a combination of pad and disc wear.' To figure out what

---

* According to Brembo, the coefficient of friction between carbon-carbon discs and pads can reach 0.9. A cast-iron disc, with standard brake pads, will have a μ of around 0.4, which is more than good enough for a road car.

proportion of the total wear is related to the pad and how much to the disc, the team combine their sensor data with historical brake data, collected in previous tests and events. 'From all that, we track our wear rate through the race. If it's too fast, we can adjust the brake balance to protect the end of the car that's experiencing the highest wear, or we ask the driver to find some clear air to cool the brakes.' Either way, the goal is to ensure that the driver has stopping power when and where they need it. With thousands and thousands of corners to face in any given season, these systems, and of course the drivers, do a remarkable job.

As Marshall and I sat drinking coffee at the end of my visit, we chatted about the relentless pace of a race season: visiting new tracks every week or two, preparing for even the most unlikely incidents, continuously tweaking systems to meet changing demands. Does the constant pressure ever get a bit much? He laughed wryly.

> F1 is always about trade-offs. You see that in all the things we talked about today – brake cooling, aerodynamic drag, downforce, tyre grip. Everything interacts. We don't have the luxury of just being able to take one thing out and optimise it. It all needs to work as a system, which is why it takes a big team of people. This sport is so challenging, but as an engineer, always getting to look at the next thing is what gets me out of bed in the morning. I love it.

CHAPTER SIX

These Shaky Isles

On 30 October 2018, I was at my desk in Wellington, New Zealand, happily writing about curling, when I felt a low rumble. For someone who hasn't grown up around them, earthquakes feel completely alien, and yet every nerve in your body seems to recognise them as a threat. We'd experienced several brief tremors since moving from London a few years before, but this was the first time I had to follow the NZ government's 'drop cover hold' advice, and clambered under my desk, phone in hand. The shaking lasted a few seconds, and did nothing more dramatic than knock some pens onto the floor. But I was still holding onto the table leg, heart racing, when my Kiwi husband phoned to check in on me a few minutes later.

This quake wasn't a serious one but it served as a reminder of New Zealand's precarious location. The country sits astride the boundary of two major tectonic plates, which are moving slowly and inexorably towards each other. The nature of that interaction depends very much on location. At the bottom of the South Island (Te Waipounamu), the Australian Plate dives, or subducts, below the Pacific Plate. Just off the east coast of the North Island (Te Ika-a-Māui), the situation is reversed – there, the Pacific Plate plunges below the Australian one. In between, there's almost no subduction. Instead, beneath most of the South Island, the two plates grind along and into each other. It's the complexity of these interactions that gives New Zealand its dramatic landscape. From the snow-tipped movie-star good looks of

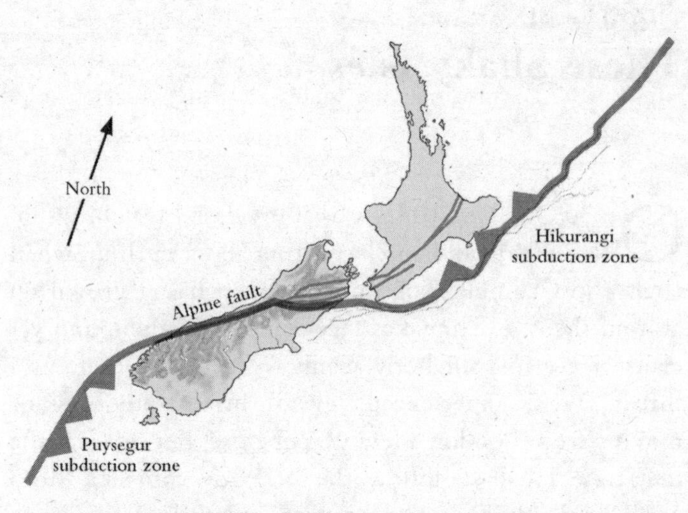

Figure 15: Aotearoa New Zealand sits on a complex and dynamic plate boundary, and faults stretch across the landscape.

the Southern Alps, to the soft sandy beaches of the Coromandel Peninsula, this country can't deny its geology.

The Pacific Plate and the Australian Plate are two of the eight(-ish) rigid blocks that the Earth's outer shell, its lithosphere, is divided into.* Driven by heat currents and enormous pressures within the rock beneath them, these vast tectonic plates move relative to one another at a rate of a few centimetres per year. While that might not sound fast, this motion has made and remade the planet over billions of years, and plate boundaries are where most of the action happens. At divergent boundaries, the spreading apart of plates creates new crust. Under the sea, this forms features like the Mid-Atlantic Ridge; on land, the result is a rift valley. Convergent and transform boundaries – like those that cross New Zealand – tend to deform existing crust, as the plates

* According to GNS Science, 'There are 7–8 major plates and many minor plates. *World Atlas* says there are 9 major plates.

move towards each other (convergent) or along each other (transform). They are responsible for the more … dramatic geological events, namely mountain-building, volcanoes and earthquakes. They're also the reason the country is known as 'the Shaky Isles'. In just one year – and I'm assured this is fairly typical – GeoNet's extensive network of seismic instruments detected 20,759 earthquakes.* While only a tiny fraction of NZ's quakes are ever large enough to be felt by humans, their existence is a part of daily life, and they have long been woven into the indigenous tales of the islands. Rūaumoko, a Māori god, is said to cause the rumbling of quakes and the hiss of volcanic activity as he moves below the earth. But as we are all too often reminded, the impact of these events is felt beyond the world of mythology.

In 2019, a tiny island in the picturesque Bay of Plenty made headlines around the world. Whakaari (White Island) erupted, claiming the lives of 22 people, and seriously injuring 25 more. The island had been a tourist destination for decades, with 10,000 visitors a year lured to its shores by the promise of low-level volcanic activity – hissing vents, deep craters and bubbling mud pools abound. Whakaari has never been truly silent. It is the tip of an enormous conical volcano that has been continuously active for at least 150,000 years. GNS Science, New Zealand's premier geoscience institute, has had permanent monitoring systems on the island since 1976. Two weeks before the fatal eruption, GNS raised the volcanic alert level, saying that patterns of signals from their detectors suggested that the island 'may be entering a period where eruptive activity is more likely than normal'. Despite this,

* Data accessed on GeoNet's statistics page on 10 February 2021. The number of events that had occurred in the preceding 365 days = 20,759.

tour operators continued to offer daily visits to the island. The result was fatal.

While Whakaari was a tragic reminder of the Earth's dynamism, the reality of living on a plate boundary is that geological unrest isn't really something you can avoid. Yes, you can assess and monitor the risks, and prepare in the best way possible, but there's always a chance that a 'big one' will come along and upend everything. As recent New Zealand history will attest, earthquakes can happen in the most unexpected places, and behave in ways that contradict all prior knowledge.

Before we can begin to tackle those particular complexities, let's cover some of the basics. The first is that the vast majority of quakes occur on faults, which are breaks in the Earth's crust.* As tectonic plates perpetually slide and crunch relative to one another, huge pressure builds up in the plates' upper layers, causing cracks that can be anywhere from tens of centimetres to many hundreds of kilometres in length. These cracks or faults mark a line along which the crust has moved. Faults also act as a zone of weakness between blocks of crust, so that any future motion will preferentially occur along them, making them of major interest to earthquake geologists.

Fault motion is usually either very slow or incredibly rapid, but it's what causes the motion that's worth discussing. Take two similarly sized pieces of modelling

---

* This is true for all but the very deepest quakes (> 600km (370 miles) below the surface). Deep quakes are still not well understood. My friend and planetary geologist at Washington University in St. Louis, Associate Professor Paul Byrne, told me that even though temperatures are very high at those depths, it's possible that parts of the mantle may be able to fracture just like the upper lithosphere.

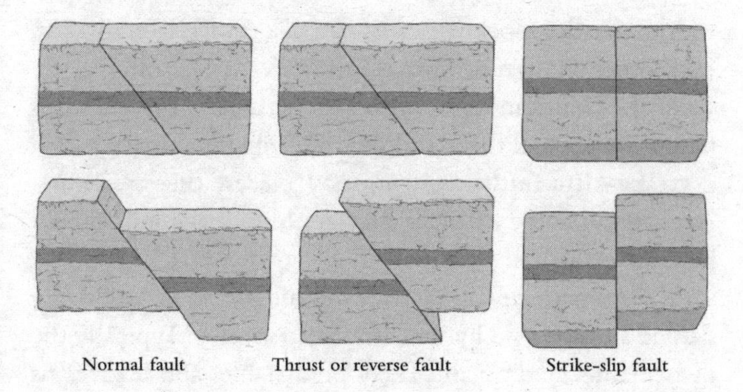

Figure 16: The three types of fault shown here can cause earthquakes.

clay and shape them into blocks. Put them side by side and bring them into contact, allowing them to slightly stick to one another. Well done, you have created a fault! Now, to apply some forces. At divergent boundaries, where the plates are moving away from each other, rocks are under tensile stress. In other words, the forces acting on them are pulling them apart. You might see your clay model stretch and deform as you tug on both ends. In the crust, additional stresses are present too, but the dominant form of stress in these boundaries is tension. The result is what's called a **normal fault**, where the rock on one side of the fault drops downwards relative to the other side.

As you might expect, at convergent boundaries, where the plates are moving towards each other, compressive stresses dominate. Because the clay is so malleable, your model will only tell part of this story. While rocks deep in the crust do tend to deform slowly, those higher in the crust respond to compression by fracturing. Typically, the result of squeezing stresses is a **thrust fault**, where during times of seismic activity, one side moves upwards relative to the other.

And finally, blocks of crust can also move laterally past each other, along transform fault lines that are often near-vertical. The dominant stress that acts on rocks near this fault is horizontal shear and the resulting structure is called a **strike-slip fault**. In your clay pieces, the stretching, twisting outcome of these stresses close to the join should be easy to see.

All along the planet's plate boundary zone, rocks are shaped and strained by these multiple stresses. Typically the deformation is slow and fairly predictable, and it gives us the iconic folded rocks we see in Namibia and the Canadian Rockies. But over time, compressive and shear stresses can build up in those rocks, and when it all gets too much, they move, releasing energy in the form of seismic waves. This sudden slippage of crust, caused by an abrupt release of accumulated stress, is what we call an earthquake.

That description of the process omits an important factor, one that helped to form the modern theory of earthquakes. It also explains why a chapter on New Zealand's geology can be found in a book about surfaces. When it comes to understanding seismic motion, **friction** rules all.

Stick

Have you ever marvelled at the crisp, clear notes produced by a concert violinist, or been spooked by a creaking door in a quiet house? If so, you're already familiar with the type of friction that dominates earthquake generation. **Stick-slip motion** is the result of a frictional instability that can happen when two touching surfaces move relative to one another. You might imagine that if both surfaces were perfectly smooth, they could slide freely – in theory, the friction between them would be non-existent. But in reality, if we zoom in far enough to discern the individual atoms that make up a surface, it becomes clear that few materials are truly 'smooth'. Even the most highly polished sheet of glass is actually covered in a densely packed array of atomic hills and valleys. In most materials, this roughness is considerably larger and it means that two surfaces sliding along one another will experience some resistance, or friction. As they slide, these tiny features can interact and even lock together, temporarily halting the motion. Static friction begins to kicks in – the longer the surfaces remain stationary, the harder it is to get them moving again. And all the while, the shear stresses that initiated the sliding motion continue to strain the material. This period is the 'stick'. Eventually, the force working to make the surfaces slide will exceed the frictional force that's holding them stationary. At that point, the surfaces will rapidly slip forward before stopping again. Assuming nothing else changes within the system, this intermittent cycle of stick – slip – stick – slip will repeat ad infinitum.

If you want to experience stick-slip for yourself, put one of your fingers onto a table (or any solid surface) in front of you, and slide it forward. Does it move easily? Now repeat the motion, but this time, do it while pushing down as hard as you can. Your finger should judder forward, alternating between moving and stuck, in a constant battle with friction.

Stick-slip frictional behaviour can occur anywhere two surfaces move against one another, and across all length scales, from atoms to tectonic plates. While there's still some debate on the underlying physical mechanism that causes it (much more on this in Chapter 9), we see its impact everywhere. For example, stick-slip is so central to a violin's sound that players regularly apply rosin, a special resin, to their bow to enhance the effect.* It is also the source of the musical notes that emerge from a wine glass when a wet finger is rubbed along its rim. The phenomenon is more often considered a nuisance or a problem to be solved – for those who operate precise mechanical systems, stick-slip is the enemy.

The first geologists to formally make the link between the jerky, stop-start motion of sliding surfaces and the mechanics of earthquakes were William F. Brace and James D. Byerlee, in the 1960s. The dominant model at the time suggested that quakes signalled a piece of crust reaching its breaking point. Namely, they occurred when the accumulation of strain was so great that it caused the rock to fracture, releasing all of that stress by forming a fault, before returning back to its original position. While it reflected seismic motion observed both in the field and in the lab, this model had some limitations. Firstly, it fell short of explaining how stress could build up in rocks that had been weakened by previous fractures. It also hugely overestimated how much the stress dropped during real, shallow-crust earthquakes. So, in 1966, Brace and Byerlee set out to understand the process, using a series of lab experiments.

They started by applying a compressive load to a cylinder of unfractured granite, gradually increasing the stress until a fault formed. When it did, it was accompanied by a sudden stress drop and a small slipping motion along the fault. Once that motion stopped, Brace and Byerlee reapplied the stress. This time, the stress drop occurred at

* The same material is used by gymnasts, dancers and drag racers – always to improve grip between surfaces.

a slightly lower level than before, and it didn't coincide with the formation of a new fault – instead, the granite merely slipped progressively further along the existing fault, before stopping again. They wrote, 'This jerky sliding could be continued almost indefinitely on the fault, with the stress building up and then being released. Each release of stress was accompanied by a small amount of slip on the fault.' (If this is beginning to sound familiar, that's a good sign.)

In their second experiment, the researchers pushed it further by using a granite sample that had been artificially fractured, cut in a similar orientation to a 'natural' fault. Again, they observed a recognisable pattern of jerky sliding, though the stresses involved were much lower. They repeated the experiments on polished and roughened rocks of varying chemical composition. In all cases, this stick-slip motion occurred. Brace and Byerlee's results allowed them to fill in some of the gaps in existing models. By showing that 'an earthquake might represent the release of only a small portion of the total stress supported by the rock', they explained why even major earthquakes occurred at stresses that rocks should have been able to withstand. It also described a possible mechanism for a sudden release of energy in rocks that were already broken and fractured.

Their model doesn't describe every idiosyncratic detail of earthquake behaviour – if it did, we'd be able to use it to predict their occurrence – but it was a huge step forward in our understanding of these events. It also highlighted the need for a better understanding of rock-friction physics, a field that, in turn, would go on to transform how we carry out large-scale infrastructure projects. These days, earthquake fault motion is almost exclusively viewed as frictional sliding on a fault plane. Virgin rocks do still fail and fracture, especially close to the surface, but the behaviour of known, active faults is dominated by stick-slip movement. On a real fault, periods of 'stick' can be decades or centuries long, whereas the 'slip' is typically just a few seconds.* With such a rapid energy release, it's not surprising that earthquakes are often compared to subterranean explosions. Sentences like 'Its power [was] on a par with … 400 atomic bombs detonating' or '[it's] the equivalent of … eight million tonnes of TNT' are fairly typical in news media, but none of the geologists I interviewed were particularly keen on the analogy. As Jon Carey from GNS said, 'Describing a quake as a detonation makes it sound like it happens at a single point, and that some sort of trigger has been pulled. The truth is much more complicated.'

Related to this is another number that usually accompanies a headline-making earthquake: its magnitude. We (non-geologists) tend to think of this as a measure of how destructive a quake is, but that's not *strictly* true, or at least it paints only part of the picture. In the early part of the twentieth century, magnitude was based on the Richter scale, which used seismometers to determine the ground motion caused by a quake and the travel distance of the quake's seismic waves. By combining these via a specific equation, geologists could come up with a metric for the size of the quake. While it worked well for smaller events, the Richter scale tended to underestimate larger earthquakes, so it gradually fell out of favour.

* In Japan's 2011 Tōhoku earthquake (and resulting tsunami), which claimed the lives of more than 20,000 people, the slip lasted for several minutes. This is part of the reason the event was so devastating.

These days, the standard used by organisations like GNS and the US Geological Survey is called the **moment magnitude**. It uses ground-motion data from highly sensitive seismometers to determine the length of the slip and, critically, the area over which the rupture occurs. It combines this with information on the ruptured rock itself – namely its shear modulus (or rigidity). This relationship gives geologists a much more reliable way to quantify the amount of energy released during a seismic event. It also means that quakes in rocks that are difficult to bend or shear tend to have the highest magnitudes. However, to understand the impact that an earthquake will have up here on the surface, we also need to know its focus; the depth at which the quake occurs.

The shaking that caused me to hide under my table on that day in October (and, it turns out, led to the NZ Parliament being temporarily suspended) was a Magnitude 6.2. No matter what scale you use, this quake would be considered a strong one. Almost 160,000 people spread right across NZ submitted 'Felt it' reports to the GeoNet web portal. And yet, it caused zero damage to infrastructure on the surface. Depending on where you are on the planet, the Earth's crust is somewhere between 5km and 50km thick (about 3–30 miles). But this earthquake's focus was far below that. GeoNet sensors measured it at a depth of 207km (129 miles), well within the mantle. As a result, the seismic waves it generated, which can travel at up to 8km per second (29,000km/h, 18,000mph), had to propagate through a huge volume of rock before reaching the surface. That had the effect of attenuating or damping the waves, dissipating their energy and reducing their potentially destructive force. Earthquakes that occur at more shallow depths release their energy relatively close to the surface, so even a lower magnitude event can cause localised damage to buildings.

So when looking at a quake's capacity to wreak havoc, we need to consider both the moment magnitude and the quake depth. The type of fault that an earthquake occurs on also makes a difference to its impact, as do both the strength of the rock involved and the area over which tectonic forces operate. For all of these reasons, subduction zone megathrust faults, where one tectonic plate dives beneath another, continuously producing compressive stresses over a huge area, are home to the largest, deepest quakes on Earth. The next largest tend to occur on transform boundaries, along strike-slip faults. Here, the release of shear stresses can displace crust by tens of metres in a single event. New Zealand's geology is dominated by faults like these – namely the Hikurangi subduction zone that follows the east coast of the North Island, and the Alpine Fault that extends along most of the South Island. These are Shaky Isles indeed.

Lab

It was a bright, sunny day when I headed to the main GNS campus just outside Wellington. Set in the expansive Hutt Valley, and sandwiched between a golf course and a large park, it's a picturesque spot. As an adopted local, though, I know the lab's secret – it sits almost atop the largest, most active fault in the region: the aptly named Wellington Fault. I still haven't decided if that makes it the perfect location for the lab, or the weirdest one. Regardless, having driven past it lots of times, I was excited finally to visit. I was mainly there to interview Dr Laura Wallace, a geophysicist I'd been gently stalking on social media ever since moving to NZ.* But in addition to talking to me about her own remarkable research (which we'll get to), Wallace had kindly introduced me to Dr Jon Carey, lead scientist of GNS's geomechanics lab. The measurement of rock and soil mechanics are his thing, making him the perfect person to direct my stupidest friction-related questions to..

'We're experimentalists, really,' Carey said, sitting opposite me in his office, wearing a shirt bedecked with tiny pelicans. 'Our job is to understand how things deform and move and slide and slip.' His group wants to observe changes as they happen, which is almost impossible to do in the field. Instead, they reproduce geological events as accurately as possible in controllable, measurable environments. A lot of Carey's experiments focus on landslides but he's also worked on samples from NZ's many faults, because, he said, 'While they might seem different, they are all part of the same kind of problem.' Ultimately both involve the deformation of rock, and a key characteristic of that process is **pore-fluid pressure**. Most rocks are made from particles of various sizes, shapes and chemical composition. But in terms of understanding how any given rock will behave, the spaces between those particles are just as important as the particles themselves. That's because in the ground, these pores tend to house fluids that can directly influence the rock's mechanical properties. You only need to think about the difference between dry and wet sand to appreciate the potential contribution of water. And while natural gases and oil can be present in porous rocks, water is the fluid usually under discussion.

---

* By 'stalking' I mean following her tweets, reading her research papers and news stories. Don't worry.

'It is a fascinating material,' Carey said. 'People think of water as a lubricant because it can make surfaces more slippery, but that's not usually how it works geologically. What makes it so influential for us is that it doesn't change volume very easily under load.' In other words, water is almost impossible to squash or compress, and in fact, pore fluids exert their own pressure, which helps a rock resist some of the many stresses acting on it. 'The water trapped inside rocks is constantly pushing back, which leads to this idea of an effective stress,' Carey explained. 'The higher the water content, the lower the effective stress acting on that rock.' He assured me that if you go down far enough into the crust, pretty much everything is saturated, and water is important closer to the surface, too. Since the 1930s, there have been countless studies that link elevated pore-fluid pressure to the triggering of landslides. And when combined with rapid shaking, its impact can be catastrophic.

In 2010 and 2011, the NZ city of Christchurch was hit by a series of earthquakes that claimed many lives, and changed the urban landscape for ever. Part of what made the events so damaging was the sandy soil on which the city sits. When the shaking started, it caused the soil to compress under gravity, closing the spaces between the grains. In a dry solid, this would increase the density of the material. But here, the water present in the pores opposed that change, and its pressure began to rise rapidly. Eventually, like the steam escaping from a boiling kettle, the pressurised water tried to find a way out, and moved upwards, carrying sand and silt with it. 'You end up with this kind of bubbling jelly that's lost all of its strength,' said Jon. 'For as long as the shaking continues, what had previously been a perfectly solid piece of ground behaves

Figure 17: This car fell into a sinkhole created by a quake in Christchurch, New Zealand. The dark grey liquefaction mud surrounded it, trapping it in place. The car's occupants escaped harm.

like a liquid.' The process of liquefaction is terrifying – one that recalls childhood nightmares of getting caught in quicksand. Photos of the immediate aftermath of the Christchurch quakes showed scores of subsided buildings and numerous vehicles trapped *in* (not *on*) roads. The longer-term legacy of liquefaction was that entire suburbs have been permanently replaced with green fields; the land was declared too risk-prone to build on.*

Effective stress is just one of a host of factors that lab-based geologists have to consider while designing their experiments. 'Every experiment has its pluses and minuses,' Carey told me, as we walked down the stairs towards the geomechanics lab, 'and no one machine can measure everything. But each one gives us a way to explore fundamental processes. Modelling helps link

* To view an interactive map of Christchurch's liquefaction zones, search for 'Canterbury Maps Liquefaction Susceptibility' on any web browser.

that knowledge to field observations, and when we tie all of those elements together, we can build a pretty good understanding of what's happening below the surface.'

Following him through a large blue door, I was faced with a room that looked more like a workshop than a scientific lab. 'This is the sample prep room,' Jon said, pointing to various crates of core samples and shelves full of rocks peppered with cylindrical holes. The space was clean and tidy, but clearly well used. Aids to help move and cut heavy samples were all in their place, with hazards clearly marked and labels aplenty. The air was filled with a musty, oddly reassuring smell. Rock dust. I liked it immediately. 'The size and shape we make each sample depends on the tests we've got planned. There's always a compromise, though,' he sighed. 'Smaller samples are easier to prepare but they give you less of the natural variability we see in the field. And if we move to larger samples, we accept that we can't get as close to the things we want to measure.' Their other main challenge is getting their hands on samples, because rather than carefully storing them for posterity, Carey's lab destroys them. 'The official term is "testing them to failure",' he chuckled. As someone who used to treat my test samples like precious gemstones, I could understand why other geologists might not be so keen to share.

Walking up a small ramp, we entered the lab itself and were greeted by a gentle hum of various motors, air-filtration systems and water pumps. All around us were pieces of equipment, seemingly one from every era of geological science, and lots of computer monitors. Lab technician Barbara Lyndsell was busy working on one of the largest bits of kit – a tall frame that supported two large steel rams, one positioned directly above the other with a large gap in between. From this, Lyndsell had removed a cylindrical vessel containing a rubber-clad rock. 'This is one of our triaxials,' said Carey, pointing at the load frame. 'The sample sits here on the pedestal, and the rubber tube stops water getting through to the sample or escaping from it.' In this case, the sample had gone through an 'undrained test'. By saturating the sample within the tube and then encasing it in a pressure-controlled water bath, Carey and Lyndsell had reproduced some of the stresses the rock would experience in the ground. And because the inputs can be so precisely controlled, this set-up offers a way to directly measure changes in the sample's pore-fluid pressure as different loads are applied.

It turns out that the ability to control the volume and/or pressure of water is an important tool in the shed of every professional geomechanicist, including Carey. 'We've always got so many variables to consider when interpreting the behaviour of a rock or soil sample. This approach helps us narrow down the scope, so that we can measure what we're really interested in.' Walking across the lab, he continued, 'For soil mechanics, especially if you're involved in civil engineering, effective stress and excess pore-fluid generation are what you mostly think about. For faults, it's all about friction and shear. We can measure shear on that load frame, but here's another option.'

In front of us was a rectangular vacuum chamber, a bit larger than a shoebox, sitting open on a heavy-duty aluminium bench. I peered inside and saw a thick steel block with a square hole. 'Is this where the rock sample sits?' I asked.

Carey nodded. 'But can you see that it's made of two blocks, one of top of the other? When the lid comes down onto the chamber, it bolts some screws into the sample holder. They're then used to lift the top block slightly.' Maintaining a tiny gap in the sample holder is important – it creates a natural weak plane in the rock. During an experiment, only the bottom half of the sample is pushed along by a shear actuator. The top half is held stationary. Once the shear forces reach a critical threshold, the rock fails, or faults, along the weak plane, producing a shear surface. This not only provides information on the strength of the sample, but if they keep on pushing, they can also look at friction on the fault. 'We have quite a lot of distance we can travel with this set-up, which means we can measure how friction evolves with displacement.'* In short, Carey can study how the fault behaves – does it slip freely, does it stop right away, or is the motion more stick-slip?

The other thing that this particular shear box can do is shake, which, he explained, is a huge bonus. 'The dynamic properties of materials can be very, very different from the static ones. So having this brings us much closer to what we see in nature.' Getting a sample ready to test is a slow process. First the chamber is flooded with carbon dioxide, to push out the air. Then, specially de-aired water is dripped slowly through the sample and fills the chamber. Once the system is completely saturated, a constant load is applied to the sample to consolidate it, echoing the rock's natural environment. Depending on the sample, and the

---

* By 'quite a lot of distance', Jon means 12–14mm (0.47–0.55in), but from a geomechanics point of view, that's more than enough to be useful.

depth they're trying to reproduce, this consolidation step alone could take a few days. Doing all that preparation for an event that's over in a couple of seconds might seem crazy, but it's worth it. By ensuring that their experiments accurately reflect real ground conditions, scientists can then trust their measurements, and know that their data is good quality. Every one of these tests – and those carried out in similar labs across the world – gets us closer to untangling the complex knot of forces that shape the ground beneath our feet.

Source

Something nagged at me when I first started working on this chapter. Given that all the rocks on a plate boundary are constantly shifting and straining in response to stress, I found myself wondering why we ever have any seismically quiet periods, and what actually makes an earthquake stop. 'Stick-slip!' I hear you yell. I know, I know, but that's the 'how'. It doesn't really explain **why** it happens. What is it about a fault that decides whether it's going to creep quietly or make a huge earthquake?

'When we feel a quake, it's really a combination of three things: what's happening at the source, what happens to the waves as they travel out from the source, and what changes near the surface. For your question, we only need to look at where the slip is happening – the source itself,' said Dr Carolyn Boulton, as we sat in a busy atrium at Victoria University of Wellington. I'd come to the right place. Boulton is an expert on the quakes of the Alpine Fault and has more than a decade of experience in measuring the frictional properties of fault rock.

She started by explaining that faults are rarely composed of bare rock surfaces. Rather, they're lined with **gouge**, a general term for the fine-grained and often claylike rock debris produced by a fault's grinding forces. As a result, the properties of gouge are central to understanding the frictional behaviour of the fault itself. When we spoke, Boulton and a group of colleagues had just published a paper in the *Journal of Structural Geology* which looked at one specific type of gouge – saponite, found in the southern sections of the Alpine Fault. More than half of all quakes along this fault are known to terminate suddenly at the boundary between its central and southern sections. So Boulton set out to explore what role saponite gouges might play in this weird, quake-stopping behaviour.

In order to understand the results of this study, we need to take a minor detour into the world of rate- and state-variable friction, which Boulton described as 'the framework with which we try to understand the frictional properties of rocks'. It sounds fancy, but in effect, it's a set of equations that help geologists interpret lab observations of rock behaviour, and there are multiple versions of them in use. There are a few key takeaways, though:

1. The *static* friction coefficient between two surfaces depends on the length of time they've been in contact with each other. So, the longer a fault has been 'stuck', the higher the friction is between its surfaces.
2. The *dynamic* friction coefficient depends on the sliding velocity, though other factors, such as temperature, also have a role to play.

3. If a fault experiences a sudden change in sliding velocity, its friction also changes value, over a specific slip length that depends on the roughness of the surfaces.

This velocity-dependence of friction is usually summarised in equations as the 'friction rate parameter' (a–b), which can be determined by experiment. Its value tells you whether a fault containing that rock will slide in a stable or an unstable manner. This is what Boulton wanted to measure in her southern Alpine Fault saponite samples, and she did it using a specialist triaxial shear system at the US Geological Survey's lab in California.

'No matter what normal stresses we applied to the saponite, it stayed frictionally very weak,' she said. 'But when we applied a large velocity step, as might happen when a wave from a distant quake hits a fault, we found it to be rate-strengthening. The value of (a–b) was positive [> 0] at all temperatures, pressures and sliding velocities.' In other words, when she gave her samples a sudden kick, they got *stronger.* 'It means that this rock is never going to nucleate [initiate] an earthquake – the slip will always be frictionally stable.' And if a quake can't get started in a specific rock type, it means that a quake that happens elsewhere can't propagate through this rock either. Its strengthening behaviour acts like a negative stress drop, denying the rupture the energy it needs and literally stopping it in its tracks. In contrast, Boulton had previously found that chlorite-bearing gouge materials from the central Alpine Fault flipped from rate-strengthening to rate-weakening as temperatures increased. 'At high temperatures, these materials displayed massive stress drops,' she explained. 'They're very unstable, which means they can nucleate an earthquake.'

Other factors are involved in a fault's behaviour during a quake, including fault orientation, the moment magnitude and pore-fluid pressure, but when asked for a simple summary, Bolton laughed and said, 'In essence, we now look at faults as comprising regions of really unstable and really stable stuff. The way a fault behaves depends on the relative distribution of that stuff.' Faults are messy – more like a poorly made patchwork quilt than a smooth cotton bedsheet. This makes it incredibly difficult to accurately describe all of their properties in just a few equations. Rate and state friction laws do a pretty good job, but a phenomenon discovered at the dawn of the twenty-first century showed that they needed an update ... and that our home planet still had a few secrets up its seismic sleeve.

Slow

For almost 20 years, the Hikurangi subduction zone has been the professional obsession of geophysicist Dr Laura Wallace. An American by birth, Wallace has spent most of her career studying the complex plate boundary processes that define NZ's largest and most active fault, and her work has fundamentally altered what we know about earthquakes. So, as I sat alongside her in her office at GNS, I racked my brain to formulate a good question, while also trying hard not to fangirl.

'OK, so what am I looking at?' I eventually asked, awkwardly, staring at her computer screen. It displayed a map of NZ's North Island overlaid by two huge blobs

– one blue and one red – bisected by a wobbly line pointing south-west to north-east.

'It basically shows that the plate motion isn't uniform,' Wallace told me. 'This blue area is steadily creeping, but the red area is where the plates are currently locked together.' It seems that the Hikurangi subduction zone is no exception to my 'faults are messy' rule. The zone marks the convergence of two plates: there, the Pacific Plate plunges under the Australian one at a rate of 32mm (1.3in) per year. At its shallowest, the trough that marks the actual interface between the plates is just 3km (1.9 miles) below sea level. That happens off the east coast of the North Island, but as you move westward, the interface sinks. 'So, here in Wellington, it's probably 25 kilometres [15 miles] or so beneath us,' Wallace explained. 'The plate dives down, deepening to the west.' In the deepest parts of subduction zones, rock is hot and squishy, which allows the plates to deform and slide along each other with relative ease. But at shallower depths, rock is more brittle, so it resists this motion. Friction dominates, causing the plates to temporarily lock together, and stress begins to build. 'A fault like this can be stuck together and accumulating stress for hundreds, or even thousands of years,' said Wallace. 'Eventually, that stress will overcome the strength of the fault. If it slips quickly, the result will be an earthquake.'

The 'stuck' portion of NZ's plate boundary is sizeable, 70km (43 miles) wide and 140km (86 miles) long, and it sits below a populous area of the country that includes the capital city, so there's an obvious need to monitor its behaviour. This is where GeoNet comes in. What started as two lowly seismometers in January 1989 has become a vast national network of geophysical instruments:

everything from sea-level gauges to air-pressure sensors. GPS receivers are particularly important in understanding ground deformation, because they allow geologists to pinpoint and then precisely track positions on the surface of the Earth.* It's these units that first identified the stuck plate boundary. They did something else too, Wallace said, smiling: 'The added value of continuous monitoring is that sometimes you see things you never thought you'd see.'

Back in 1999, a Canadian geophysicist called Herb Dragert noticed something weird about a series of GPS stations he was monitoring on Vancouver Island, home to the Cascadia subduction zone. He wrote that 'a cluster of seven sites briefly reversed their direction of motion,' with some moving backwards, relative to the plate's usual course, by as much as 4mm (0.16in). Typically, a slip of this length would accompany a fairly big earthquake, but this event seemed to be devoid of any seismic signal. Crucially too, this 4mm slip happened over a period of 15 days, rather than just a few seconds. Later that same year, a similar finding was made by a group of researchers working on a different boundary, the Bungo Channel in south-west Japan. Dubbed **slow-slip events (SSEs)**, they were a geological oddity, sitting somewhere between the steady slow creep of plate motion and the rapid slip of an earthquake. They *do* release energy – often as much as a large quake – but they do it so slowly,

* Technically, GPS is the name of the United States' Global Navigation Satellite System, which was the first one in operation. But in a similar way to the word 'hoover' becoming interchangeable with all vacuum cleaners, so too has GPS with all satellite navigation systems.

and so benignly, that they're almost undetectable. And in science, any time there's something that doesn't fit existing models, there's a group of researchers eager to study it.

'It was a really exciting time,' said Wallace, sitting back in her chair. 'I'd arrived in May 2002, specifically to help design the new continuous GPS network for GeoNet. With all these results from Cascadia and Japan, we were desperate to get our units set up in the field.' So desperate in fact, that Laura voluntarily spent her first weekend in NZ setting up a GPS station in Gisborne, a large town on the east coast. It was worth it. 'In October 2002, I noticed this weird eastward movement at Gisborne, of a couple of centimetres over two weeks. I thought, "Holy crap, we've got slow-slip too!" And in terms of surface displacement, our events were way bigger than what had been previously reported.'

Since then, geologists have developed a much greater understanding of slow-slip events, which are often called silent earthquakes. Firstly, they occur in most of the major subduction zones on the planet, and even a few transform faults, like the San Andreas Fault in California. Secondly, according to Wallace, they seem to sit on a feather edge, 'straddling a transition between rate-strengthening and rate-weakening behaviour'. Thirdly, SSEs are not truly silent. Imagine listening to an audio recording of someone speaking quietly into a microphone in a noisy room; anything potentially interesting they have to say is being swamped by the general cacophony. Now imagine that instead of one microphone, you have ten, all trained on this person's voice. If you played all of those recordings at once, you might be able to find common areas in the audio signals – maybe not specific words, but you could figure out when the person is speaking. That's the basis on which scientists discovered 'the chatter of silent slip', officially called **episodic** or **non-volcanic tremor**. One seismometer is not enough to determine that anything out of the ordinary is happening, but by comparing the traces from multiple seismometers, it's possible to identify patterns; to pick out the signal from among the noise. In a now famous paper, Dragert and his colleague Garry Rogers directly compared several years of GPS displacement data in the Cascadia zone with the tremor activity recorded on seismographs. There was a clear correlation between the two, suggesting that while these slow-slip events aren't accompanied by 'traditional' seismic waves, they do generate 'unique non-earthquake seismic signatures'.

Jeremy Gosselin from the University of Ottawa told me that on the Cascadia boundary, slow-slip events occur 'like clockwork, every 14.5 months', but the same cannot be said for SSEs on the Hikurangi subduction zone. The dozens of events Wallace and her colleagues have detected have proven to be a diverse bunch. In the northern part of the subduction zone, SSEs tend to be 'shallow [< 15km, 9 miles], short [< 1 month], and frequent [every 1–2 years]', whereas in the southern section, they occur at greater depths, last up to a year and are less common. And finding conclusive evidence for tremor in NZ's slow-slip events has also proved to be a challenge. While it does seem to occur in some events, in others, SSEs are accompanied by more typical seismic signals. This might just reflect the Hikurangi margin's complex structure, but it also doesn't help that most of the trench is buried under kilometres of water. How do you pick out a tiny, noisy signal in the relentless ocean?

'Hobbits?' I asked, certain I must have misheard Wallace. 'I know we're in New Zealand, but … really?!'

She laughed, and said, 'This one is spelled H-O-B-I-T-S-S, and stands for Hikurangi Ocean Bottom Investigation of Tremor and Slow Slip.' The HOBITSS experiment, which she led back in 2014, deployed 39 instruments on the seafloor off Gisborne. Of these, 24 were super-sensitive seafloor pressure gauges, which continuously measure the absolute pressure exerted by the overlying water column. Wallace explained the basic idea behind them: 'If a slow-slip event happens and it causes the sea floor to rise – even by just a few centimetres – that means the sensor has less water above it, so it records a pressure drop. If the sea floor drops, you measure a pressure increase.' Just a few months after deploying them, the array of gauges detected a large slow-slip event near the Hikurangi trench, making HOBITSS the first successful demonstration of this technology in the world. 'Our GPS network is amazing, but it only tells us what's happening on land,' said Wallace. 'With the addition of the pressure sensors, we can better pinpoint the timing of slow-slip events. Being closer to where they happen means we're getting a much clearer picture of our seismic landscape.'

One thing that geologists are still struggling to explain is the *cause* of these silent quakes. Rock type and frictional properties are one option. The other is pore fluid. 'For a few years, there's been an idea that slow-slip events happen in areas with really high fluid pressure,' said GNS's Dr Emily Warren-Smith. 'That would reduce the normal stress, potentially making it easier to slip.' But, she said, that doesn't explain everything, 'They're episodic events, which suggests that something must change in order for them to happen.' Warren-Smith set out to search for clues in data from the HOBITSS ocean-bottom seismometers. 'We noticed a change in the faulting style before and after a slow-slip event. And the stress field changed too.' She continued, 'Combined, these observations point to an increase in fluid pressure being the driving force of the process.' Warren-Smith and her colleagues now believe that fluid released from the subducting plate gradually builds up on its surface and lubricates it. To them, it's this slipperiness that triggers a slow-slip event, which can last weeks, months or years. Deformations and cracks formed during the slip allow the fluids to drain, reducing the pressure. Eventually minerals will fill up the cracks, resealing the system, and allowing the fluid pressure to increase once more. And so the cycle continues.

A Cascadia subduction zone study that used entirely different measurement techniques came to very similar conclusions. Gosselin, lead author on that paper, said, 'During slow-slip events, we see changes in seismic velocity that could only reasonably be explained by fluctuations in pore-fluid pressure.' So, it's looking promising.

As I reached the end of my interview with Laura Wallace, I found myself pondering the broader implications of slow-slip events – can they trigger larger-scale quakes? Wallace replied,

> We know that they're a huge player in accommodating the motion between the Pacific and Australian plates in NZ's North Island. And while we've seen hundreds, even thousands, of slow-slip events around the world, only a few of those have led to really big earthquakes. But as I see it, understanding this relationship in space and time between slow-slip events and earthquakes is one of the most important things that we can do. Because if we can

understand that interplay, maybe one day we'll be able to do a better job at forecasting bigger earthquakes.

Shock

The Kaikōura peninsula, on the north-east of NZ's South Island, is famed as a haven for marine life. Visitors from all over the world come to catch a glimpse of the sperm wales, dusky dolphins, orca, seals and albatross that occupy its shores. But on 14 November 2016, it made news headlines for a different reason. Just after midnight, a shallow M7.8 earthquake struck near the small town of Waiau. In just 74 seconds, it had punched its way north-east, 'unzipping' along a 170km (105-mile) stretch of the coast. In some areas, land was horizontally displaced by as much as 12m (40ft), while elsewhere, huge landslides led to the closure of the state highway and main railway line. Two people tragically lost their lives.

This event is now considered one of the most complex earthquakes ever recorded on land. It generated the strongest ground acceleration of any New Zealand quake, and jumped between faults separated by 15–20km (9–12 miles). This challenged a long-held assumption that a gap of just 5km (3 miles) between faults would usually be enough to stop an earthquake. It also activated more than 24 faults across two distinct seismotectonic systems, while somehow bypassing the Hope Fault, the 'source of the largest seismic hazard in the region'. In the immediate aftermath of the Kaikōura quake, geologists rushed to gather data – everything from collecting soil and rock samples and digging trenches across faults, to tracking displacement via GPS, aerial imaging and LIDAR (Light Detection and Ranging, using lasers to measure distance).

Since then, nearly 750 papers have been published on the 2016 event, each one hoping to unravel some of the mystery.*

One, from researchers at Germany's Ludwig Maximilian University, suggested that the non-rupture of the Hope Fault was due to 'unfavourable distribution of dynamic stresses'. Another study, led by Université Côte d'Azur, tried to link the Kaikōura event to a 'loading of the Wairarapa fault' in the North Island. Dr Wallace and her team observed multiple, weeks-long slow-slip events – all in the North Island – directly following the quake. They also believe that the plate boundary beneath the upper South Island slipped by half a metre in three months following the event, 'slowly releasing energy … equivalent to a magnitude 7.3 earthquake'.

'The whole thing was surprising,' said Dr Rob Langridge, another of GNS's earthquake geologists. 'But even within what was a really unusual sequence, the rupture of the Papatea Fault was especially dramatic.' Langridge has been mapping NZ's active faults for 20 years, so I wasn't particularly surprised to see a huge number of graphs, photos and maps adorning his office wall. He led me to one in particular, which showed all of the known faults north of Kaikōura. The Papatea Fault immediately stood out because it was oriented so differently from the others around it. Rather than trending north-east, it curved southwards. 'The Papatea Fault had previously been mapped as a geological fault, but there was not enough documented evidence of faulting on young surfaces,'

---

* Google Scholar search on 3/3/20 for 'Kaikoura quake', excluding citations and patents = 748 results. Interestingly, when I searched for 'Kaikōura quake' (with the macron included in the place name – which is correct), just 243 results showed up.

Langridge said. In short, it wasn't considered to be an active fault. And yet, during the event, he said it produced 'the largest vertical movements of all the faults that ruptured. The fault and the triangular block associated with it moved upwards by 8–10 metres [26–33ft], and south by 4–6 metres [13-20ft], depending on where you are.'

For a fault to move that much, it would be reasonable to assume that there must have been a significant build-up of stress, right? 'That's what we concluded in a paper in mid-2018,' said Langridge, chuckling. 'I think I said something like "For a fault to move 8–10 metres, it'd need 5,000–10,000 years to accumulate strain". Well, once we trenched the fault we found that it's actually had four movements in the last thousand years. So, that was unexpected.'

Researchers at the University of Victoria, Canada, got in touch with Langridge, eager to study the unusual behaviour of the fault. They fed data from airborne LIDAR surveys of the region (carried out before and after the earthquake) into a new algorithm they'd developed to allow them to map the surface displacement in 3D. What they found surprised everyone, even Langridge. They concluded that the Papatea Fault slipped without releasing inter-seismic strain energy. Instead, they wrote, it 'occurred in rapid response to shortening caused by neighbouring elastic ruptures'. So rather than rupture to release built-up stress, the Papatea Fault failed because it was squeezed by the movement of other faults around it. Their slip 'activated' this fault, triggering it to slip, too.

By any measure, this is weird behaviour, and it flies in the face of some fundamental assumptions of fault mechanics. And from a practical, 'living safely in seismic zones' point of view, the rupture of the Papatea Fault also has important and widespread implications. While it may never be possible to accurately predict the time, location and magnitude of a future earthquake, geologists can and do develop short- and long-term hazard forecasts based on probability. These take into account factors like previous seismic activity, patterns of behaviour and knowledge of ground conditions, and they're widely used by governments, insurance providers and engineers to make planning decisions. Earthquake forecasting is also usually based on the 'elastic strain cycle model', which says that faults gradually accumulate strain until they suddenly fail, releasing energy. But the Papatea Fault did not follow this model. It lacked the key ingredient – stored seismic strain – but it ruptured anyway. And it wasn't just any slip. It produced multi-metre displacements that, if they had occurred in a high-population area, could have had a catastrophic effect. For these researchers, such results show that 'we cannot rely on strain accumulation rates alone as an indicator for rupture potential.'

★★★

Since the earliest days of geology, scientists have learned a huge amount about the processes that shape and mould our world. We're so familiar with them that we can look at structures on other planets and describe how they formed. But the ground beneath our feet is far from stable. From fractures and the ever-changing force of friction, to fluid pressures that ebb and flow, seismic activity is still full of surprises, and it's possible that we might never truly understand it. But that's not to say we shouldn't try. The geologists featured in this chapter are among the most driven and most engaging scientists I've met. They're in it for the long haul, determined to pick through gigabytes of data, test every sample they can and spend their weekends installing new sensors. All in an effort to keep us safe. As I sit here at my desk, in a house literally built on sand, very close to an active fault, that thought gives me some quiet comfort. But I've got my earthquake emergency kit at the ready anyway…

CHAPTER SEVEN

Break the Ice

Why is ice slippery? It might sound like a question that a five-year-old would ask, but it's a surprisingly difficult one to answer.

In school textbooks, ice's slipperiness is usually, if incorrectly, attributed to something called pressure melting. The idea is that when you put pressure on ice – say, directly under the thin steel blade of an ice skate – you create a lubricating layer of water that provides the slip. The act of applying pressure to a solid pushes its atoms closer together, making it denser. And because water is denser than ice, when ice is under pressure, some of the solid turns into liquid water.* The resulting layer is so thin that it refreezes as soon as the blade has passed. No one doubts that water is involved in skating – one awkward fall on an ice-rink will give you an appreciation of the general wetness of ice. And given that it's possible for a wire to pass through a block of ice using nothing more than two weights, pressure certainly plays a part too. But that's far from the whole story.

Research has repeatedly shown that the sharpest of blades, worn by the heaviest of skaters, could decrease ice's melting temperature by no more than a couple of degrees. *Maybe* that effect could be enough to melt ice that is relatively 'warm'; so at or around freezing point,

* We all instinctively know that ice is less dense than water – it's why a lake can have a layer of ice that floats on top of the water beneath – but it is still weird. But then ice is a very strange sort of solid.

0°C. But the optimum temperature for various winter sports is far below 0°C. Professor Bob Rosenberg, reporting on earlier work, suggested that figure-skating ice should be between -3 and -5.5°C, while for hockey, around -9°C is the ideal. And over the decades, multiple Arctic explorers have reported being able to ski at -30°C. Pressure melting can't explain the presence of liquid water at these frigid temperatures. Retired Professor of Physics Hans van Leeuwen recently described the idea that ice could be pressure melted in the millisecond that a skater spends on a certain spot as 'inconceivable'. And the final death knell for pressure melting? It doesn't explain why you can slip on ice while wearing flat-bottomed shoes, which exert considerably less pressure than an ice skate.

So, something else must be going on.

'Friction!', I hear you yell. 'That produces heat, which melts the ice!' Well, let's take a look at that theory. In 1939, two researchers constructed an ice cave close to Jungfraujoch, a high-altitude research station in Switzerland. There, aided by liquid air and solid carbon dioxide, they explored the possible causes of ice melting at sub-zero temperatures (down to -140°C). Rather than skates, they looked at skis made from different materials. The idea was that, if pressure was the main culprit, then they'd see the same melting, regardless of the material the skis were made from. If friction was to blame, they'd see a difference between melting under smooth brass skis versus wooden ones.

In fact, they found a bit of both. At very low temperatures, frictional melting was found to completely dominate, which (partially) explains why it's possible to skate or ski at well below 0°C. But as you approach the melting point of ice, friction gradually decreases, and pressure melting begins to play an extremely limited role. Multiple studies have since confirmed and formalised these findings, but most still failed to answer an important question: why can ice be slippery even if you're standing still? If you're as inelegant as me, you'll know that it's possible to slip the very second your foot touches ice – and that's just not enough time to generate the friction or focused pressure necessary to melt ice into water.

If the slipperiness of ice can't be explained by external forces acting on it, perhaps it could be something intrinsic to the ice itself. That was certainly the opinion of legendary experimental scientist, Michael Faraday. In 1850, he carried out a series of experiments on pairs of ice cubes, which showed that they would freeze together when brought into contact. He called this process regelation. Faraday suggested that there was a thin film of liquid water naturally present on the surface of each ice cube, and that its presence was critical in helping two ice cubes to fuse to each other. After a lukewarm reception from his peers, Faraday published further work on regelation, which pointed to the same conclusion. But again, it was largely ignored, possibly because the existence of atoms and molecules was still under question at that time.

It would be almost a hundred years before another scientist would explore this idea of an intrinsic liquid film, but this time it stuck. Charles Gurney, a Professor of Engineering at Cambridge, proposed that molecules on the surface of the ice were less stable than those deep within the ice because they had few other water molecules to bind to. And so, he said, the motion of these unstable molecules is what prompts the formation of a layer of liquid water. Gurney's paper really opened the floodgates, and as new experimental tools and techniques have been developed, we've learned more about what we now call surface melting, or premelting: the development of a 'liquid' layer on the surfaces of solids far below their official melting point. For example, we now know that these layers don't just occur on ice – they've been found on many other solids (a discovery which I'll admit leaves me with a slightly icky 'ewww, everything is a bit wet' feeling). We also know that their thickness – estimated to be somewhere between 1 and 100nm – depends on the temperature and the presence of impurities (such as salt) in the ice. In addition, in 2016, Japanese scientists managed to see these ultra-thin layers using optical microscopes in a lab environment, though we haven't yet managed to see them 'on the rink'.

A group of researchers from the University of Amsterdam and Germany's Max Planck Institute may well have come closest to unpicking the complex physics of ice's slipperiness. In 2018, the team, led by Professors Daniel and Mischa Bonn, were specifically looking at sliding friction – otherwise known as kinetic friction – which applies when two surfaces slide across each other, like a skate across a rink.* But what interested them most was what was happening at the molecular level. Could they figure out what individual water molecules were doing on the surface of ice? Step 1 of answering that question was to design a series of experiments in which they could controllably slide an

---

* Mischa and Daniel Bonn are brothers. We first met Daniel in the Introduction. There, he was leading a study into the sliding friction of dry, damp and wet sand.

indenter – a peppercorn-sized ball made from steel and a second one of glass – across different ice surfaces.* In their set-up, they could vary the sliding speed and the applied force, and, importantly, the temperature of the ice itself.

At the frigid temperature of -100°C, they discovered something extremely strange. When ice is cold enough, it is not slippery at all; in fact, it's a very high-friction surface. Under the steel ball, ice behaved just like rough glass. But as the temperature of the ice was increased, friction *decreased* steadily, and reached a minimum at -7°C. As they turned up the temperature beyond that, towards a balmy 0°C, they then saw a steep *increase* in friction. The same trend occurred when using the glass ball, too. This suggested that, regardless of the material, there is an optimal temperature for sliding around on ice, -7°C. Outside of that, skaters and skiers would have to contend with higher friction. The fact that ice changes its behaviour at different temperatures is probably not that surprising – just think about biting into chocolate stored in the fridge versus chocolate kept at room temperature. But the size of that change was significant. The friction coefficient of steel-on-ice was *fifty times* higher at -100°C than it was at -10°C. And then there was this mysterious transition point. Why did friction drop between the minimum temperature and -7°C, only to climb back up again as it warmed beyond that?

* I measured the diameters of 20 separate dried peppercorns for this analogy, and got an average of 4.95mm (0.2in). The indenter had a diameter of 4.8mm (0.19in). That's close enough, I reckon.

There were some clues: when scratching the ice between -7 and 0°C, the researchers could see the indenter begin to plough into the ice, visibly deforming it. At temperatures below -7°C, the indenter had no visible effect on the surface. So, whatever was happening, it was related to the strength of the bonds between molecules on the ice surface. The Bonn brothers moved on to the next stage of their research, using their data and computer simulations of individual molecules to build a computational model of the surface. That model revealed that there are two different types of H_2O on the surface of every ice sheet.

On Earth, ice tends to take on a highly regular crystalline structure.* This means that each water molecule is surrounded by, and chemically bonded to, four neighbours. H_2O molecules on the surface of ice, however, tend to bond only to three neighbours, which is still enough to hold them in place. But what Mischa and Daniel found was that, mixed in among these triply bonded molecules, there were some that were bonded to only two other molecules. That difference – three bonds versus two – has a huge impact. The Bonn brothers found that the doubly bonded molecules were highly mobile, and could roll around the surface of the ice like tiny ball bearings.

'Their remarkable mobility really surprised us,' Daniel Bonn told me over the phone from Amsterdam. Their other unexpected finding was that, as the ratio of mobile to

* Despite crystalline ice being by far the most common form of ice on Earth, amorphous ice dominates in our Solar System and beyond. It forms at incredibly low temperatures and can be found in comets and on icy moons, among other locations.

stationary molecules changed with temperature, it perfectly corresponded to the change in friction they'd measured in their experiments.* The more mobile the water molecules, the lower the friction. At -100°C, almost all of the surface molecules are tightly bound to the rest of the ice, producing a hard, high-friction surface that scratching has little impact on, and that's impossible to skate on. But rising temperatures increase the energy of the surface molecules, allowing some of them to loosen their bonds, gradually decreasing the measured friction. By the time the ice reaches -7°C, the majority of molecules on the surface are now mobile, making it slippery, while the bulk of the ice retains its characteristic hardness.

Remember that this all happens at temperatures below the freezing point of water, which is why, during our conversation, Daniel was so adamant that the mobile molecules should not be referred to as 'water'. He even went as far as to say that their study 'finally showed Mischa [his brother, co-author and a chemist] that there's no layer of water at the surface of ice'. I guess technically he is correct – at anything below 0°C, water really should be called 'ice' because it is a solid. But the image of weakly bound molecules freely rolling around on a surface definitely makes me think of it as a liquid.

'So what else should we call a layer of highly mobile ice molecules?' I asked.

---

* This is a relative term – atoms are never really stationary. At all temperatures above absolute zero, they jiggle constantly. As temperatures increase, so too does the level of jiggling. Eventually, they jiggle so much, they can shrug off their bonds completely, turning a solid into a liquid, or a liquid into a gas.

'OK, maybe I'd go as far as calling it a quasi-liquid,' he responded, laughing. The high mobility of these molecules has another beneficial side-effect for skaters. At the sweet-spot temperature of -7°C, any scratches or damage on the surface of ice are continuously filled by the quasi-liquid, smoothing the surface almost instantly. The result is the perfect low-friction surface that is a pleasure to glide across.*

Above this temperature, it's the hardness of the ice, rather than its rolling surface molecules, that defines its behaviour. Between -7 and 0°C, the bulk of the ice gradually softens, and sliding objects begin to dig into it, rather than move across its surface. This change from a material that can recover from deformation to one that can't is the cause of the increase of friction on 'warm' ice observed by Bonn and others.

So it seems ice is slippery thanks to a combination of several effects. One relates to the presence of liquid-like molecules on its surface, in a layer that, even below freezing, behaves differently from the bulk. It's also related to just how mobile these surface molecules are, because up to a point (or at least up to -7°C) the warmer ice gets, the lower its frictional characteristics. And finally, as we approach its melting point, the decreasing hardness of ice plays an important role on just how easy it is to skate across.

Most people can go through life without suffering the consequences of ice's complex slipperiness. But every four years, ice surfaces hit the news headlines all over the world. And as the Winter Olympics remind us, not all ice is created equal.

---

* It's likely that this same molecular mobility contributes to the healing of ice that we see in weighted-wire-passing-through-ice experiments Derek from Veritasium has a nice video about regelation on his YouTube channel. Search for "Ice Cutting Experiment" to find it.

Skate

Ninety-two teams competed in the 2018 Games in PyeongChang, South Korea. The US had the largest contingent, followed by Canada, Russia and Switzerland. The 22nd largest team at the Games didn't have any flag-bearers though, because they weren't athletes. They were the 'ice meisters'. They had the unenviable task of ensuring that the ice, used by eight different sports across multiple arenas, behaved exactly as expected for 16 days. The ice meisters are the guys (and they are almost always guys) who drive onto the rink in big machines during a break in the action, to slowly and methodically resurface the ice.* But in addition to that critical maintenance, they put months of effort into laying exactly the right kind of ice on each rink and track, with every sport having its own specific requirements.

Everything starts with the water, which needs to be as close to pure H_2O as they can get it. Depending on where you are, the ice cubes in your freezer may contain traces of nitrogen from the air, fluorine, salts and lots of other dissolved minerals. While such impurities make no difference to our nicely chilled drink, they're a nightmare for the ice meisters, because they subtly alter the molecular structure of water and, therefore, the properties of the ice.† To avoid this, they only use water that has been highly filtered, way beyond drinking-quality standards, and which contains zero impurities and as little air as possible. Then, there's the design of the rink or skating oval, which is usually very simple – a concrete slab with a

---

* There was one female driver at the 2018 Winter Olympics: Barbara Bogner from Colorado.

† The impurities in water provide its flavour. Ultra-pure water tastes of nothing.

network of pipes embedded beneath the surface. By flowing a cold, salty solution through these pipes, the concrete can be cooled to -9°C, allowing any ice laid on top to stay frozen.

Speed skaters – who regularly reach speeds in excess of 50km/h (31mph) on a 400m-long (437yd) oval track – need to move as quickly as possible on the straight parts of the track. They also want to power off the start line, and safely speed around the corners, so they're in a constant battle between glide and grip. Skaters can minimise how much contact they have with the ice through the design of the skates – their polished steel blades are just 1mm (0.04in) thick. And they can increase their power output through strength training and good technique. But what really determines their maximum speed is the make-up of the surface itself, which is why long-track speed skating demands the coldest, hardest ice of all Olympic sports.

To create a speed-skating rink, ice meisters spray paper-thin layers of their ultrapure water on the track, allowing each one to freeze fully before applying the next. Lane markers and the start/finish line are usually painted on after four or five layers of ice have been applied. More ice is then applied on top of those markings. Once the ice has reached a thickness of 2.5cm (1in), resurfacing machines smooth it by drawing a sharp blade across the ice surface, before laying down a film of hot water that rapidly freezes.*

* A small aside: There's a term called the Mpemba effect, which suggests that hot water can freeze faster than cold water. To date there's been no definitive proof that this effect is real, but that hasn't stopped people asserting that it is. A 2016 paper from Cambridge physicists concluded 'somewhat sadly, that there is no evidence to support meaningful observations of the Mpemba effect'.

It can take two weeks to ice a speed-skating oval, but the result is a solid, slippery surface that sits at … can you guess what temperature? Yep, -7°C, the optimal temperature identified by the Bonn brothers.

You might be wondering why these researchers needed to prove something that the ice meisters have clearly known about for ever. It's really because knowing or observing something is several leaps away from understanding it. And that's the thing about many of the surfaces we're exploring in this book. It's not that we know nothing about them – just like speed-skating ice or asphalt on roads, we've been manipulating and designing surfaces to work for us for decades. But surprisingly often, and usually because of the 'it works, so why worry' attitude, we haven't taken that final step: to understand the complex physics and chemistry of surfaces, and why they behave as they do. And while purely practical knowledge can take us a very long way, it has limitations, especially when we start playing around at the edges of it. Developing a deeper fundamental understanding is the key to improving almost anything.

In saying that, the skill and expertise of ice meisters is unmatched, and seems to imbue them with an instinct for their material of choice. Speaking on NPR back in 2018, Remy Boehler said that he uses his ears to determine if they've made 'good ice': 'When I arrive [at the rink], I take off my cap so I can really listen to the sound of the blades, if they're gliding or cracking.' And many skaters say they know instantly if the ice they're on is going to be fast or slow. Long track speed skaters set eight Olympic records in the 2018 Games, so the ice meisters working in the Gangneung Oval were obviously doing something right.

Next in the ice hierarchy is ice hockey, which is usually played on ice between -5 and -7.5°C. This ice is gradually laid down over four to five days. It has a similar thickness to speed-skating ice, but is slightly softer to allow the players to perform the agile turns and explosive changes of pace that characterise the game. Next, you have short-track speed skating and figure skating, which, despite having different needs, often have to share the same rink in the interest of space-saving. As you might expect, short-track skaters need relatively cold, fast ice, but because they travel short distances between corners, they need a little more grip (and therefore slightly softer ice) than long-track speed skaters. In contrast, figure skaters have to be able to dig the tips of their blades into the ice in order to launch into their spins and jumps, and land safely. So they need a much warmer ice (-3°C) than any of the other competitors, and ideally much thicker layers of it too. It takes about three hours to switch between these two types of ice. Usually, it's easiest to start with thin short-track ice, and once the competition finishes, ice meisters can gradually increase the temperature and add more layers, to get it ready for figure skating. If they need to switch back, resurfacing machines carefully shave the ice while the rink temperature is reduced.

As you can imagine, keeping the ice at the ideal temperature is no mean feat, either. Another ice meister, Mark Callan, told me that in Korea, 'The average temperature outside was -8°C but inside the arena, at head height, the air temperature was +10°C. If you're not careful, that can cause havoc with the ice. Three thousand people walking into the area carry a lot of heat inside – it's our job to manage that.' Depending on the sport, emergency repairs can be carried out by flooding the ice, or filling large ruts with slush or snow that is rapidly frozen using pressurised CO_2. These masters of ice know all the tricks.

While the slipperiness of ice plays an undeniably large role in all winter sports, I can only think of one sport in which athletes can actively manipulate its slippery nature during the competition.

Curl

The Scottish game of curling was first introduced as an Olympic sport in 1998.* Like millions of others, every four years since, I've found myself glued to this bizarre scientific ballet of granite, humans and brooms. Curling has a long history. The earliest known curling stone, found in a pond in Stirling, central Scotland, bears the year 1511 on its surface, and Robert Burns wrote about the game in 1786 in two famous poems – 'The Vision', and 'Tam Samson's Elegy'. Curling involves two teams sliding polished granite stones along a 46m-long (50yd) rectangular sheet of ice. The aim for each team is to position their stones as closely as possible to the 'bullseye' (more correctly called the button) of a target painted just beneath the surface of the ice.

The sport gets its name from the curved path taken by these massive stones, which, according to the World Curling Federation, must weigh no more than 19.96kg (44lb). The curl is initiated by the player rotating the stone as they release it to slide down the sheet; a non-rotating stone will not curl. Two other players, called sweepers, accompany the thrown stone on its journey. Using a flat-faced broom, they furiously sweep the ice in front of the moving stone, and straighten its path. This makes curling the only sport in which the direction of the projectile can be changed after it leaves a

* Curling had also been a demonstration sport at the 1932, 1988 and 1992 Games.

player's hand, which is pretty cool. But if I left the description of curling at that, I'd be selling it short. Physics is involved in all sports, controlling everything from the top speed of a cyclist to the spin of a cricket ball. But with curling, physics is embedded in every moment of action in the game. Despite this, we still can't say for sure what causes the stone to curl the way it does.

Let's start with what we do know. First, the stones used in the Olympics are very special indeed. Shaped somewhat like a squashed satsuma, they're made from two types of granite found on Ailsa Craig, a 5km-wide (3-mile) island off the coast of Scotland. The bulk of the stone is made from common green granite, which is tough and slightly speckled. For the 'running surface' of the stone – that's the layer that makes direct contact with the ice – manufacturers turn to blue hone granite, a fine-grained rock that is unique to Ailsa Craig. Thanks to their mineral structure, which results from their volcanic past, these granites are extremely durable, non-porous and shatter-resistant. They are also highly water-repellent; especially blue hone. This stops ice forming on the stone's surface, which reduces the risk of freezing-related damage that might compromise the stone's performance. Something that you may be surprised to learn is that the underside of a curling stone is not flat. In fact, this running surface is concave, a bit like the bottom of a beer bottle. As a result, only a narrow ring of blue hone granite (~6mm, 0.2in wide) actually makes contact with the ice.

Now, the ice itself. If you've ever watched curling on TV, you might have noticed that curling ice lacks the high-gloss shine of the speed-skating oval. That's because it's not smooth – while the underlying ice is flat to within one-hundredth of a millimetre, it's then deliberately covered in tiny ice bumps, called pebbles. Specialist ice meisters walk up and down the

Figure 18: Curling is a curious ballet of stones, brooms, and ice.

rink, spraying water droplets through different-sized nozzles onto the ice, until it is uniformly pebbled. Generally, two layers of pebbles are applied to a curling sheet – one in each direction – to ensure that the ice retains its roughness for the entirety of the game. The ice meister then draws a blade across the entire sheet, limiting the maximum height of the pebbles. Pebbling reduces the contact area between the stone and the ice, which in turn reduces the friction between them. Its pebbled slipperiness is unusual, but it is essential to the sport. Curling stones move very differently on smooth ice.

And finally, the brooms. The pad that touches the ice is made with a specified, but fairly unremarkable, coated nylon*. Sweeping immediately in front of the stone melts the ice, producing a layer of lubricating water ('real' water this time, not the quasi-liquid from earlier) that reduces the friction between the surfaces. This helps the stone travel further, and it straightens out the curl on the stone's path. All of the curlers I spoke to suggested that the faster and harder a sweeper can sweep, the straighter and further a curling stone will travel.

* Oxford 420D, in case you're interested.

These three components – the pebbled ice, granite stone and brooms – combined with the skill of the players, are what give us the game of curling. So far, so good. The controversy really only arises when we start talking physics, so let's do an experiment.

Take an empty beer bottle or an upturned glass and put it on a smooth table or countertop. What we're trying to do here is emulate the ring-shaped running band of a curling stone. Very carefully, set the bottle sliding along the table in a straight line. Where does it end up? Now do it again, but this time, make the bottle spin a little bit, so that it rotates as it slides. Does it end up in the same place as before? It turns out that the spinning bottle will always veer away from the straight path that the non-spinning one took. If you rotated the bottle to the right (clockwise), it should end up curling to the left. If you rotated it to the left (anti-clockwise), your bottle curls to the right.

This happens because of the way friction acts on the bottle base. From the moment a sliding bottle leaves your hand, friction works to slow it down. As it decelerates, the bottle will lean forward slightly, pushing the leading edge into the table, increasing the friction there.* If the bottle is spinning as it slides, the leading edge will always experience more friction – and so, move slower – than the back edge. So the motion of the faster moving back edge is what dominates the direction of travel; in short, if you spin it to the right, it curls to the left. This is called **asymmetric friction** and, in physics terms, it is entirely unsurprising. However, a curling stone flouts these rules. It will curl in the *same* direction in which it spins – so, if a player sets it

* This friction also explains why a sliding object will always tip over in the direction of motion.

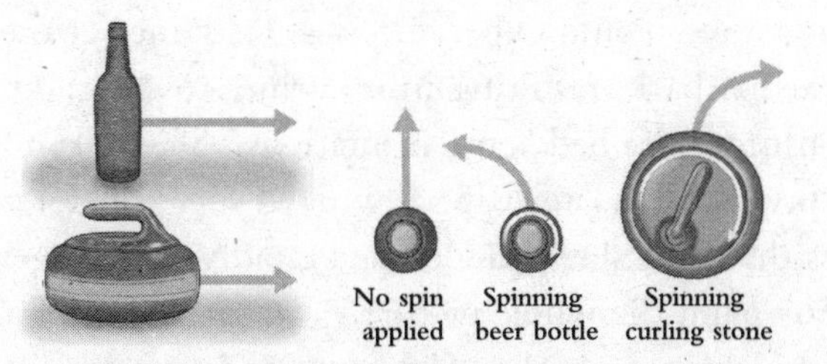

Figure 19: Curling stones are still full of surprises — their motion doesn't really follow any of the 'obvious' rules of sliding objects.

spinning to the right, the stone will also curl to the right. The reason that it does this is, perhaps unbelievably, still under debate. As of mid-2020, physicists working on this question could be grouped into two camps, and each one thinks the other is completely wrong. We have ourselves an old-fashioned science face-off.*

By day, Mark Shegelski is a Professor Emeritus of Physics at the University of Northern British Columbia in Canada. By night, he is a recreational curler, and started publishing scientific papers on curling in the 1990s. Back then, he proposed a 'thin liquid film model' for the motion of a curling stone, which had some similarities to our bottle-on-table experiment. The idea was that, just like the bottle, the curling stone tipped forward as it slowed, which applied pressure to the leading edge. But rather than it causing the leading edge to slow down, Shegelski suggested that this increased pressure would warm the ice, producing a thin film of water that would actually *reduce* the friction at the front of the stone. So, unlike a bottle, a clockwise-spinning

* Opposing views are very common in science – when you come up with an idea, the onus is on you to test it rigorously. If it's a new or controversial idea, you expect other groups to question it. This is at the heart of the process called peer-review, which aims to ensure the quality and integrity of published science.

curling stone would experience less resistance at the front than at the back, resulting in it curling to the right. 'The thin-film model had some limitations,' Shegelski told me on a Skype call in late 2018. 'It worked very well for curling stones that slide slowly and rotate rapidly. But I knew that it didn't paint the whole picture.'

Shegelski's original model remained popular for years – despite its limitations, it described much of the odd sliding-curling behaviour of a curling stone. And none of the proposed alternatives offered a better explanation. That changed in 2013, when Staffan Jacobson and his colleagues at Uppsala University in Sweden published two papers pointing to an entirely different mechanism. Uppsala's 'scratch-guiding' idea was based on experiments they'd carried out at a local curling rink. Their results suggested that the stone's running band was rough enough to leave microscopic scratches on the surface of the pebbled ice. In their version of events, these scratches would be laid down by the leading edge of the curling stone, and when the back edge encountered them – because don't forget, the stone is rotating – it would get a tiny jolt, and move ever so slightly to follow those scratches. Jacobson, speaking to me in early 2019, said, 'The first scratches act as a sort of track to guide the protrusions on the rough surface.' These pushes might be microscopic in size, but because of the roughness of the stone and of the ice, they occur many times over, and result in the stone curling in the direction of rotation. This, the authors said, removed the need for a thin film of water – in other words, Shegelski was wrong.

Jacobson and his colleagues took the idea further by replacing brooms with harsh sandpaper, sweeping the ice in specified directions to deliberately scratch it.

Videos they produced at the time showed the stone veering left and right in accordance with the scratch direction.* Broomgate, a 2015 controversy that rocked the curling world, seemed to give the Jacobson team further confidence that they had the right idea. A company had launched a curling broom with a 'directional' fabric on its brush pad, and it quickly became clear that it could change the game. When in the hands of elite sweepers, these 'frankenbrooms' could be used to actively steer the stone, changing its path so much that it could completely compensate for a poor throw. Emma Miskew, an Olympic medal-winning curler, was quoted as saying, 'You really shouldn't be able to steer a rock down the sheet. That's not curling.' With lots of unhappy teams protesting against the use of these brooms, the World Curling Federation commissioned a study. It revealed that the frankenbrooms didn't just polish the ice – they heavily scratched it, and that by changing their technique, sweepers could indeed use those scratches to guide the stone, even to go against the original curl. As a result, these brooms were banned from major competitions.

Despite this, Shegelski just didn't buy scratch-guiding. 'A good idea is one thing, but to develop a theory, you need quantitative results,' he told me. 'These guys don't have any calculations that accurately describe the motion of the rock, so to me, they don't have a theory.' When asked what he thought of their sandpaper experiments, Shegelski said, 'The basis of their idea is that the curling stone scratches

---

* Destin from SmarterEveryDay has an excellent curling explainer video on his YouTube channel: Search for "Cold hard science. The Controversial Physics of Curling" to find it.

the ice as it slides. So, why use sandpaper at all? Just scratch the ice with the stones. If the idea stacks up, we should see wiggly paths of curling stones, veering left and right, in every game. And we don't.'

Roughness certainly plays a role in the interaction between ice and stone. After all, a stone with a highly polished running band won't curl, even on pebbled ice, and curling-stone manufacturers are very protective of their roughening methods. What Shegelski and the Uppsala team disagree on is exactly what the role of roughness is. 'Broomgate didn't prove their idea at all,' said Shegelski. 'Yes, the brooms scratched the ice, but they're so much deeper than anything a stone could create. Even if a scratch-based guiding mechanism might apply in extreme conditions, it tells us nothing about a standard, real-life curling game that uses non-abrasive brooms.'

Shegelski and his colleagues stated this (and more) in letters published in response to the Jacobson paper. That, in turn, prompted a written reply from the Jacobson team. This is the scientific equivalent of declaring a duel. At the time of writing, the teams had never actually spoken – their only communication channel to date had been through the medium of research papers. I pointed out the weirdness of this to Jacobson. He was fairly magnanimous, saying, 'I've never met Shegelski, but I'm not interested in having a spat with him. We just showed that his model wasn't enough to explain the observations.'

Shegelski, when asked if a collaboration would offer a better outcome, said, 'When I first heard their idea, I sat down and tried to build it into existing models, and to compare it to our observations over the years. I hoped it would work, I wanted it to, but it absolutely didn't. Our model might not have been complete, but theirs was just incorrect.'

For Shegelski, the most mysterious question of curling has to do with the rate of the stone's rotation, which, he says, the Jacobson papers do not address. To understand it, let's go back to our bottle-on-table experiment. We established earlier that the bottle will 'curl' if you spin it as you release it. There is also a well-established link between the rotational speed of the bottle and the lateral distance it will travel, which means that if you spin the bottle faster, it will curl more. This is *not* the case with curling stones. Observations from several groups have shown that it doesn't matter how many times a stone rotates on the ice, the curl distance stays the same, at around 1m (3.3ft). 'It is bizarre,' Shegelski told me, 'and it has long caused problems for those of us trying to understand the physics of curling.' Neither Shegelski's original thin-film model nor Jacobson's scratch-guiding mechanism could fully explain this observation.

Then, in 2015, Shegelski got a phone call from Dr Edward Lozowski, a physicist from the University of Alberta. 'Ed is usually referred to as the "king of ice physics" here in Canada,' said Shegelski. Lozowski had previously published papers on speed skating and bobsledding, but woke one day to find himself thinking about curling, and knew that Mark was the man to talk to. That call kick-started a collaboration between the two, and a year later, they published their 'pivot-slide' model that they believe offered the most complete description of the motion of a curling stone to date.

The theory is that, as the stone slides and rotates, the microscopic roughness on its running band causes it to *briefly* stick to the ice pebbles on the surface. This contact time really is brief – Shegelski and Lozowski estimate it to be about 45 nanoseconds, ten million times quicker than the blink of an eye. During that time, they say, the stone pivots about the pebble in the same direction as the rotation, pulling at it until it loses contact. The stone then continues to slide, albeit in a slightly different direction, until it encounters another pebble, which causes another pivot. Because this will happen tens of thousands of times as the stone slides down the pebbled ice sheet, it could cause sufficient redirection to explain the mysterious curl of a curling stone. Shegelski and Lozowski continued to develop their model, and in 2018, published a new equation of motion for a curling stone.

$$x_F = \frac{32}{3\pi}\frac{r}{\sqrt{\frac{1}{2}R^2 + r^2}}\frac{H}{E}\left[\frac{L}{d}\right]^3 h\,N\,{v_0}^2{t_F}^2$$

Final curl distance, in metres

Properties of the stone

Properties of the ice and the pebbles – controlled by the sweepers and the sliding stone

The speed of the rock – controlled by the thrower

If you're not a fan of maths, this equation might look daunting, but stay with me. On the left-hand side you have x_F, which is the curl distance. On the right, there is a number ($32/3\pi$), which is related to the shape of the stone, followed by another stone-specific term (*R*= radius of the stone, *r*= radius of the running band). The next group of terms relate to the ice. Some are its mechanical properties – *H* is its hardness, and *E* is its Young's modulus,* likened to the 'elasticity' of the ice. Others focus on the ice pebbles (*L*/*d* is the height of the pebbles divided by their diameter,

* Young's modulus is a measure of the stiffness of an elastic material: it tells you how a material will respond to being compressed or stretched in one direction. While this property is named after polymath Thomas Young (1773–1829), he was not the first to use it. Leonhard Euler (1707–1783), an incredibly influential mathematician of the 1700s, published a paper describing the same property 80 years before Young. So maybe we should call it Euler's modulus?

N is the number of pebbles in a particular area, and *h* is the height reduction in the pebbles once a curling stone has slid over them. These ice-related terms are what the sweepers alter – by rubbing their broom on the ice in front of the stone, they soften it (reducing *H*) and melt some of the pebbles (reducing *L*). This would make the curl length shorter – in other words, sweepers make the stone curl less, which is exactly what we see in real life.* The second-to-last term (v_0) is the stone speed, which is down to the skill of the thrower. And t_F is the stopping time, which is largely defined by the friction between the ice and rock. It tells us that on perfectly smooth ice, the stone won't curl ($x_F \approx 0$), which has been backed up by multiple observations.

Almost as interesting as what's included in this equation is what's missing from it. Shegelski and Lozowski found that, for a slowly rotating stone (typical in a game of curling), the total curl distance (x_F) is *independent* of the rotational speed of the stone. So, it doesn't matter how fast it's spinning – the curling stone will always curl by the same amount. And by plugging typical values into their equation, they get x_F = 0.93m (3.05ft), which is comfortingly close to the frequently observed curl distance of about 1m. This is a big point in its favour. Shegelski also contends that 'broomgate' – which Jacobson and the Uppsala team presented as proof of their competing theory – actually supports *his* pivot-slide mechanism. To him, 'Scratching up the ice would increase the number of points that the rough running band of the curling stone can pivot on.' When we spoke, he was in the early stages of testing this idea on the rink.

So did that mean they'd come up with a 'Grand Unified Theory of Curling', I asked? Shegelski, laughing, answered,

* Wheelchair curlers don't sweep, so their shots need to be even more precise than those of able-bodied curlers.

'Maybe we have, or if not, then at least we've gotten pretty close!' And after our conversation, I found myself agreeing with him. Although it's far from perfect, Shegelski's pivot-slide model makes a sort of instinctive sense – it *feels* right because it reproduces many of the real-world observations of curling, and it is based on solid physics. But while the Jacobson paper had its flaws, and required a few too many leaps of logic to connect cause and effect, I couldn't fully let go of the scratch-guiding idea. As the scribbles in my notebook from the time attest, most of my doubts would be answered if the Uppsala team developed a reliable mathematical model, and did some more detailed experiments.

It turned out that experiments were already on their way. In late 2019, I came across a paper from researchers at Finland's Aalto University. Though unconnected to Jacobson, the team had set out to test his scratch-guiding theory for themselves. They started by building a custom, 3m-long (9.8ft) curling rink inside a 'cold room' (read: a freezer big enough to house a lab). What made it special was that an A4-sized section of the pebbled ice track could be quickly removed and placed under a microscope. This allowed the Aalto team to scan the ice surface before and immediately after a curling stone slid over it. If there were any scratches present, they could then measure the angle between them, and compare that to the scratch-guiding predictions. At speeds typical of competitive curling, the researchers observed 'cross-scratches caused by the leading and trailing edges of the linearly moving and rotating stone', and they found a strong correlation between the scratch angles and the sideways displacement of the stone. In other words, their results supported the idea that scratches produced by the curling stone could guide the stone's path.

Separately, a Canadian researcher called A. Raymond Penner published his own mathematical model of scratch-guiding, based on Jacobson's initial paper. Scratch-guiding was fighting back. True to form, Lozowski and Shegelski prepared a formal response, describing the Penner model as 'a welcome potential explanation of how the scratches made by "frankenbrooms" may alter the trajectories of curling rocks', but they questioned many of the details. The duo referred to their own experimental results, published within months of Penner's paper, and which Shegelski had told me about during our earlier interview. They concluded that 'the lateral deflection of curling rocks cannot solely be due to scratches made by curling rocks'. Within a few weeks, Penner had penned his own reply to these comments, largely disagreeing with them. Could this be the start of another beautiful rivalry? Who knows!

For now, the world of curling remains divided into scratches versus pivots, but the complete answer to the question 'what makes a curling stone curl?' is likely to be some combination of these effects. Given how much research has been published on the topic in the past five years, it's perfectly possible that by the time you're reading this book, there'll be another dominant theory. That's just how science works, and I had to stop somewhere. The thing I love most about curling is that this charmingly odd Olympic sport, first played on frozen Scottish lakes 500 years ago, still has its secrets.

Glacial

Until now, we've talked about the link between the properties of ice, and how things move on its surface. But that's only half of the story of ice's slipperiness. Let's take a look at how ice moves *on other things*, and let's do it on a grand scale.

There's no more spectacular form of ice than that found at the poles or high up in the Earth's mountainous regions. Glaciers are a literal force of nature. They store three-quarters of the world's freshwater, feed rivers, irrigate crops and sculpt the landscape beneath them. But the ice that they consist of is rather different from any 'normal' form of ice. A glacier begins with snowfall. Over time, as thick layers of fluffy snow accumulate, they begin to densify, and those snowflakes deep within the proto-glacier melt and reform under pressure, packing ever more tightly together. In the process, air is gradually squeezed out from the ice crystals, and snowflakes start to lose their hexagonal structure, becoming more granular as a result. Season after season, this granular snow hardens, and its crystals grow, until it becomes something called firn.* After decades of this cycle of snowfall, compaction, melting and recrystallisation, some of the ice that you're left with – typically found towards the bottom of the glacier – will be so devoid of air bubbles that it appears transparent, with a blue tint. Because of the changes that this ice undergoes through its lifetime, some people class it as a type of metamorphic rock.

'I would ascribe to that,' chuckled geophysicist Professor Christina Hulbe. 'You could even consider glacier ice as a rock that's really close to its melt temperature; so, in a way it's like lava.' Professor Hulbe, from the University of Otago, has been studying glaciers and polar landscapes for close to 30 years, so when I wanted to understand how enormous blocks of super-dense ice could move, she was my first port of call. For the most part, she said, glacial motion is down to a combination of two forces: internal deformation of the ice, and sliding at its base. The relative importance of each force depends on the glacier, and no

---

* The word *firn* comes from Old High German and means 'old' or 'last year's'.

two glaciers are the same. 'In the Alps, a glacier might be on a rough, rocky surface, whereas in the Antarctic, you can have ice floating on water,' said Hulbe. 'The processes acting on these glaciers will clearly be very different – in one, you'll have lots of shearing at the base; in the other you'll have none. They're just the two end members of the glacier scale – there's a whole spectrum of possibilities in between.' Deformation does seem to be a common factor, though, and it's because when ice deep within a glacier is put under pressure over long periods of time, it will slowly and permanently distort – it is effectively a plastic. If this happens on a downward slope, gravity acting on the glacier's huge mass is enough to drive that deformation, stretching its ice crystals and making it 'flow'. In contrast, ice in the upper layers of a glacier tends to be brittle, so when it deforms, it cracks, producing huge crevasses that carve the glacier into pieces.

In some cases, the weight of a glacier is also enough to melt the ice along its base, forming a lubricating layer that will help it to slide. If there's a lot of this meltwater present, a glacier can move at a rapid pace. In 2012, Greenland's Jakobshavn Glacier was found to be moving 46.6 metres (51 yards) every day – four times faster than it did in the mid-1990s. In the Antarctic, there are also features called ice streams, which are fast-flowing channels embedded in the ice sheet. As Hulbe explained, glaciers often lie on water-saturated sediment. 'This sediment is soft compared to the overlying ice, so it deforms first. That means that there's almost no traction at the base – the ice in the stream can slide and stretch and shear. But the inter-stream ridges experience higher friction; they're effectively frozen to the bottom.'

Studies on the many processes that operate in and under the Antarctic ice are playing an important role in understanding the long-term implications of climate change. As Hulbe described in a TedX talk, recorded at Scott Base research facility, the ice sheet in West Antarctica is one of the largest sources of sea-level rise on the planet, as well as the major source of uncertainty in climate projections.* She tells me:

> We know that increasing atmospheric CO_2 is causing the atmosphere and ocean to warm, and that glaciers are responding to that through melting and changing the rate at which they flow. We know a lot about the underlying physics, and we can create computer models to study what could happen in the future. The key for us is knowing when, not if, the climate warming generates a real instability – a runaway retreat that won't stop even if the warming does. The ice sheet in West Antarctica is going to change no matter what we do next, but the decisions that we make as global communities can limit how fast it happens. There's always an opportunity to change the pathway that we're on, and the sooner we get started, the better.

Formation

Before we finish with all things ice, there's one last thing to discuss – the processes that allow ice to stick to, and form on, everything from skating rinks to airport runways.

Ice tends to form when it's cold – obvious, I know, but low temperatures alone are not usually enough to turn water into ice. In fact, under the right conditions, water

* To watch Prof. Hulbe's TEDx talk, search for "Putting the brakes on runaway ice sheet retreat in Antarctica" on YouTube.

can stay liquid down to -20°C.* There's a simple experiment that you can do at home that shows this. Take a sealed, plastic bottle of water, and place it into your freezer. Domestic freezers operate at close to -18°C, so within about two hours, the water in your bottle should be 'supercooled', meaning it will still be liquid, despite being well below its freezing temperature.† Slowly and carefully remove the bottle from the freezer. And then, when you're ready, give the bottle a short, sharp smack – I like to slam mine onto a table. If you've cooled the water sufficiently, this action should make it turn into ice in front of your eyes.

What's happening is something called **nucleation**. The energy you percussively provided to the bottle encouraged a small number of supercooled water molecules to rapidly align. The resulting structure is, in effect, an ice crystal: a 'nucleus' for other molecules to cling to. What you see as the ice 'spreading' through the water bottle is the alignment of more and more of the water molecules, as the liquid crystallises into ice. Nucleation is what kick-starts the freezing process, but it doesn't always need a jolt of energy to occur. In everyday life, it's more likely to be the result of an impurity, say a speck of dust, or the presence of a scratch or rough spot on a surface that cold water is in contact with – anything that gives water molecules a place to grip onto and grow into an ice crystal. Once you have all

---

* This is the 'practical' limit. In theory, moisture can exist as a supercooled liquid down to -40°C.

† You might need to experiment with this – the composition of tap water varies a lot, depending on your location. And your freezer settings can have an impact on how long the water takes to become supercooled. If possible, after 1.5 hours, you should check on your bottle every 10 minutes.

three – water, sub-zero temperatures and several nucleation sites – you're very likely to get ice.

'Actually, that's not quite the whole story,' Professor Amy Betz from Kansas State University told me over Skype. Missing from my description was something vital: the water vapour that's ever-present in the air we breathe. I suspect I could dedicate an entire chapter to water vapour, but for our purposes, we need to know that water in the atmosphere is constantly changing between its gaseous, liquid and solid forms, evaporating and condensing at different rates. At a specific temperature known as the dew point, the balance shifts towards condensation, which is when we see the formation of liquid droplets.* Betz continued, 'For freezing to occur, you need temperatures cold enough to be below both the freezing point and the dew point. At standard atmospheric pressure, these temperatures are usually similar, but at higher or lower pressures, there can be a gap between them. Frost, which comes from the condensation and then freezing of water vapour, won't form unless a surface is below both of those temperatures.'

And from a surfaces point of view, there are several other things that affect the formation of frost and ice: 'In addition to the humidity, the chemistry and structure of the surface also significantly impact how and when something freezes,' said Betz. In other words, how ice sticks to surfaces depends on exactly what that surface is made from. This is something that Betz knows a lot about – for several years, she has led a research group exploring the interactions between surfaces, water vapour and freezing temperatures. 'I actually started at the other end of the temperature scale,' she said. 'For my

* For more on this, search for 'Alistair B. Fraser's Bad Clouds' guide in any web browser.

PhD, I looked at the boiling of liquids on novel surfaces.' When Betz became an Assistant Professor, she wanted to change tack. 'I thought it could be interesting to look at how these surfaces control other phase changes, like freezing. The mechanism turned out to be completely different to what we'd expected. Ice is really weird,' she laughed.

Hydrophobic surfaces (like those we met in Chapter 1) have long been known to delay the growth of frost. This makes a sort of instinctive sense – after all, if water can't stick to a surface, ice can't form on it. But according to Betz, eventually, 'under certain temperatures and relative humidity, frost formation is inevitable. Delaying it is the best we can do'. When that frost does finally form on hydrophobic surfaces, it's in a less dense layer than forms on hydrophilic (water-loving) surfaces. Betz hypothesised that **biphilic** surfaces, which combine hydrophobic and hydrophilic areas, could be even better at suppressing freezing. 'Our idea was that the water vapour droplets would form preferentially on the hydrophilic areas, but the hydrophobic areas would stop them from coalescing to form a layer. Instead, they'd be pinned in place, and isolated from each other.'

This finding allowed Betz to control where and how quickly liquid could freeze on surfaces. By altering the chemistry and the shapes of features on surfaces, the team could accelerate freezing or slow it down. Betz applied for a patent on one particular surface, which was patterned so that it became covered in nanoscale pillars. She also demonstrated the same behaviour on lots of different materials, uncovering other mysteries along the way. 'For example, the ice that forms on our silica nanopillar surfaces is cubic crystalline – not the usual hexagonal ice that defines the shape of snowflakes. We don't understand why that's the case, but it's fascinating.'

Betz's work could also have practical implications. 'There are lots of industries that struggle with frost formation,' she said. 'Refrigeration, air conditioning and transport – they all have to deal with frequent defrosting, and it comes at a major cost.' One example is the aviation sector. In winter, it's very typical for frost and ice to form on an aircraft fuselage overnight or while waiting at the gate. If you've ever sat on a plane while a heated orange fluid is sprayed all over it, you might be familiar with one of the current solutions. These de-icing compounds, usually made from propylene glycol, work to decrease the freezing point of water, stopping ice from forming even at temperatures approaching -45°C. They're effective and cheap, which is why they're so widely used, but if used incorrectly, they can act as an environmental pollutant.* A chemically similar, but generally much thicker green fluid can also act as an anti-icing coating. It absorbs water and prevents it from sticking to the surface, but it is only temporary. As the plane speeds along the runway, it gradually sheds its slimy green skin, and by the time it reaches 300m (1,000ft), it has shrugged it off entirely. As you might imagine, airports go through huge quantities of these fluids. An EPA report from 2012 estimated that every year, US airports use 95 million litres (21 million gallons) of de-icing fluids. I asked Professor Betz if her coatings could change that. 'Right now, our biphilic coatings just aren't cost-competitive with de-icing fluids,' she said. 'But if there's ever a change in the regulations, or if the cost of the chemicals drastically increased, people might become

---

* Airports are bound by strict regulations around the use of de-icing compounds, and on their disposal, storage and re-use. The exact details of these regulations vary by regions.

much more interested in other options. You never know what turn technology is going to take.'

Ice formation on the fuselage isn't just a niche issue – it has a major impact on the aircraft's performance. Thomas Ratvasky, a NASA engineer, told *Scientific American* that 'Ice reshapes the surface of the lift-producing parts of the airplane: the wings and the tail. That roughness is enough to change the aerodynamics.' It increases drag, adds weight to the plane and can reduce thrust. All in all, ice is not good news. In fact, two researchers at Washington, D.C.'s Office of Aviation Safety reported that the icing of aeroplane surfaces caused 583 aviation accidents – leading to 819 fatalities – in the US between 1982 and 2000.

These numbers include incidents caused by the ice that forms on an aircraft when it's in flight, but at high altitudes, it's not snow or sleet that pilots have to worry about. The only thing that really sticks to a moving aircraft is supercooled liquid water found in certain types of clouds. Here, the aircraft acts as the sudden jolt we gave our bottle of water earlier. As it moves through a cloud, the leading edges of the plane's wings are bombarded by supercooled water droplets, which freeze and form ice structures. This usually happens at air temperatures between -10 and +2°C, with droplets on the 'warmer' end of the scale producing hazardous, horn-shaped structures called glaze ice. Glaze is notorious for disrupting airflow around the wings, which significantly affects the plane's aerodynamic performance. As the temperature drops further, the supercooled water tends to freeze on impact, producing wedges of an opaque, brittle ice called rime that can increase the weight of an aircraft.

None of the coatings applied to a plane on the ground offer protection from in-flight icing, so other systems are needed. Most large commercial aircraft route hot air through ducting in some of the plane's critical areas, such as wings and tail surfaces. Because the heat is generated simply by running the engine, this 'bleed air' acts as a continuously generated thermal barrier against icing. Heat is also the key for electro-thermal systems used on planes like the Boeing 787. They work by bonding wires to the inner surface of the wing's leading edge. As an electric current is passed through them, they heat up, which either stops the ice from forming in the first place, or causes any existing ice to melt. A lot of smaller planes take a more mechanical approach to ice removal. Black rubber membranes, called 'de-icing boots', are installed along the wing's leading edges. Compressed air repeatedly inflates and deflates the boots, breaking any ice that had formed on the surface. While all of these systems work reliably, scientists and engineers continue to search for alternatives, many based on novel, anti-icing surface coatings. Between 2015 and 2020, more than 4,500 papers and patents were published on the topic, with slippery silicone, oily graphene and even 'natural' antifreeze molecules extracted from mushrooms all proposed as ways to keep aircraft surfaces free of ice. So, will future aircraft defeat ice accretion by being super-slippery? At this point, it's hard to tell. But as someone obsessed with surfaces, I admit that I hope they will.

The slipperiness of ice sits at the heart of so many experiences in daily life, but it also incorporates some of the most interesting questions I came across in researching this book. I thought I understood a lot about ice's properties, but I was wrong. I never imagined I'd find so much controversy in a centuries-old sport, nor did I fully appreciate the mechanics of glacial motion. I don't know about you, but I'll never look at the material in the same way again.

CHAPTER EIGHT

The Human Touch

The coolness of a floor beneath bare feet. The clickety-clack of a keyboard and the buzz of your smartphone. The crisp texture of a summer top or the fluffy cosiness of a wool blanket. The softness of a loved one's skin. The reassuring hand-squeeze they give you when you're nervous. We live in an intensely tactile world, and we navigate it through the wonder material that is our skin. Its complex network of embedded receptors is constantly at work, taking cues from the environment and responding to them. Our ability to *feel* shapes us. Touch also plays a huge role in human development, and as neuroscientist Professor David J. Linden and others have shown, it can act as a social glue to bind people together.* Despite this, touch may well be the most underappreciated sense. It's so central to our daily lives that it has disappeared from view … or should I say, reach. Unlike the 'big four' senses – taste, hearing, sight and smell – touch isn't served by any particular art form.† We don't see 'tactile taprooms' sandwiched between restaurants, concert halls, cinemas and perfumeries. The closest equivalent I can think of is my local fabric store,

---

* For much, much more on what touch means to humans, I recommend the 2015 book *Touch*, by David J. Linden (Viking). He is an excellent guide to the complexity (and biology) of this sense.

† In reality, all experts agree that there are many more than five senses, though there's little agreement on exactly how many more.

but even the wide variety of textures it houses represents only a tiny fraction of the touchy-feely interactions we are capable of experiencing.

Every square centimetre of our skin is wired for touch, and each touch-related sensation follows its own distinct path from skin to brain. Pain, pressure, vibrations, stretching, changes in temperature, etc. are all felt by specific sensors. But the information they collect is blended seamlessly, and combined with information from our other sensory organs, to paint a sophisticated tactile picture of our immediate surroundings. Skin is the ultimate interface: the complex, multi-layered surface at which our inner workings meet the outside world. And while touch is considered a biological system – one whose scope stretches far beyond the bounds of this chapter – it has physical and engineering implications too. They're what I want to explore.

But let's be even more specific, because not all skin is equally sensitive. Skin can broadly be separated into two types: hairy and hairless. What you may not realise is that close to 90 per cent of your skin fits into the former category. There's the obviously hairy bits, like the scalp and legs, but even the apparently smooth skin on the inside of our forearm and bicep are covered in 'peach fuzz': tiny, soft, unpigmented hairs. Truly hairless, officially known as **glabrous**, skin can be found in just a few locations – your hands (palms and fingers), the soles of your feet, your lips, nipples and parts of your genitals.*

This glabrous skin has sensing superpowers, specialising in the type of touch that allows us to rapidly discriminate

---

* Some types of glabrous skin, e.g. lips, foreskin and vagina, can also be described as mucocutaneous because in addition to its hairlessness, it also features a mucus membrane.

between objects, and to recognise textures, surfaces and shapes. For humans, most tactile exploration happens via the movement of our hands, and they're capable of incredibly fine perception. In 2013, a group of Swedish researchers concluded that, using their index fingers, humans could discern surface features just 13 nanometres in size, equivalent to about five strands of DNA lined up side-by-side. That's much, much smaller than anything we can see with the naked eye.* And a 2021 study from the University of Delaware showed that our fingertips can distinguish between smooth surfaces that are identical, save for a single atom substitution. That's why, in this chapter, we'll be focusing on our hairless human hands, and on the many ways they allow us to interact with the world around us. Hold on tight!

Prints

I've had a few passports and visas over the years, so I'm fairly familiar with getting my fingerprints scanned. But the process felt a little different this time around. For a start, I was sitting in a windowless room at New Zealand Police Headquarters in Wellington. The image was also being collected on a device not much larger than my smartphone. And I was obsessively staring at the complex patterns in a way I never had before. Fingerprint Officer Gilane Khalil was taking me on a tour of my fingertips:

> The dark lines are raised areas, what we call papillary ridges, and their patterns are grouped into three very

---

* According to *Science Focus* magazine, humans with normal eyesight can resolve lines separated by gaps of about 0.026mm (or 26,000nm), assuming they're 15cm (6in) from your face.

> broad categories: loops, whorls and arches. In a loop, the ridges come along, loop over and then recurve, going back to the same side. An arch comes in one side, curves or tents upwards, and goes out the other – they're only found in about 5 per cent of the population. And whorls are circular patterns. Most people have a mixture of loops and whorls, and that's what you have too.

Pointing to the print from the middle finger on my right hand, she said, 'You do have an unusual one though. Can you see that this single feature is a combination of loop and whorl? That's a nice example of composite pattern.'

Though fingerprints get most of the attention, the entirety of our palms are actually covered in this papillary ridge skin, where it's overlaid with a network of characteristic deep creases and wrinkles. All of this complexity reflects what's happening below the visible surface. The ridge patterns on our skin are formed from different varieties of the protein keratin, with the strongest, most durable varieties found in the raised ridges, and more pliable forms in the valleys between them. This combination means that ridges can withstand a lot of compression, while the valleys allow them to flex and stretch.

The roots of these patterns are deep, extending below the outermost layer of the skin (the epidermis), and into the underlying dermis. This layer of connective tissue has a similarly ridged form – David Linden described it as 'inward facing fingerprints' – and it provides the epidermis with a range of supports, including blood vessels. The skin's sweat glands and ducts also anchor these layers together, feeding the vast number of sweat pores that follow along the top of the ridge-lines. The glands found beneath the

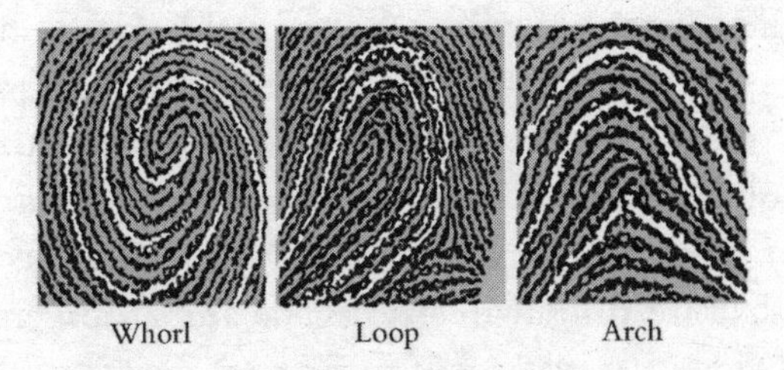

Figure 20: Papillary ridge patterns are complex, but these are the three most common. Which ones do you have?

glabrous skin of your palms are the largest and most densely packed of anywhere on the human body, with 1,000–1,200 of them in every square centimetre.* So you can blame them the next time you get sweaty hands at an inconvenient moment.

Humans are not the only primates to have papillary ridge skin on their hands and feet. A study carried out by Pittsburgh Zoo and the FBI in 2011 involved collecting the prints of various primates during routine veterinary examinations. Unsurprisingly, those species known to be closely related to us, such as orangutans, gorillas and chimpanzees, all displayed similar loops, whorls and arches, though in somewhat different distributions to our own. Almost half of all orangutan prints featured arches that are rare in humans. Chimps had many more whorls than we do, and gorillas had

* That's equivalent to 2,500–3,000 in every square inch (1 square inch is approximately 6.45cm^2). In addition, there is a widespread idea that fingers also produce sebum, the natural waxy substance that makes your hair greasy and your skin oily. But in fact, there are no sebaceous glands on your palms, nor on the soles of your feet. So, any sebum found in fingerprints has been transferred there by touching other parts of the skin with your fingers.

about the same proportion of loops as the average human. There is at least one other animal known to have fingerprints, despite having taken a very different evolutionary path to primates: the Koala (*Phascolarctos cinereus*). This fluffy marsupial (and Australian icon) has ridges that are similar in size, shape and arrangement to human ridges, with arches occurring on every digit of their front paws and on some of their back paws. Whorls and loops tend to be found only on specific digits. The general consensus for why such different species would share these specific features is that the presence of ridges improves their ability to grasp. This is a useful skill for any species that spends a lot of time in forest canopies ... or, as we'll discover later, one that's more typically found in the urban jungle.

Fingerprints have long been used by humans to make their mark on everything from contracts and clay tablets to the walls of ancient tombs. But their adoption as a means to identify individuals – due to their apparent uniqueness – is more recent, and it has a very chequered past. The three people most associated with its early development are physician Henry Faulds, eugenicist Francis Galton and colonial police officer Edward Henry. Faulds experimentally determined that fingerprints are permanent – they can recover the same patterns even after serious surface skin damage. And other than increasing in size, they stay the same from birth to adulthood. He also designed the first formal classification system for the patterns. Galton built on those assertions in an 1892 book, and after collecting examples of prints from all over the world, declared friction ridges to be 'an incomparably surer criterion of identity than any other bodily feature'. This opened the door to their use as an identification tool.

Galton emphasised the potential importance of this technique for the British colonies, 'where the features of natives are distinguished with difficulty'. Yes, he really wrote that.* Based in India and working as Inspector General of the Bengal District Police, Edward Henry embraced Galton's work, and became convinced that he could make the classification system even more practical. The result of his efforts, the Henry System, was adopted by Scotland Yard in 1901, and versions of it have been used by law enforcement and other authorities ever since.

In recent decades, the position of fingerprinting in forensic science – especially its use as a form of evidence in criminal cases – has come under criticism from a number of prestigious scientific organisations. The crux of the matter is around the concept of individualisation; that a forensic trace (say, a latent print found at a crime scene) could be unambiguously connected to one specific individual, 'to the exclusion of all others'. In 2009, the US National Research Council published a major study into the state of US forensics. In it they said that fingerprinting lacked the scientific basis required to make such a claim. Later reports agreed, pointing to issues such as error rates, a lack of repeatability and

---

* In line with many of the other racist beliefs he held, Galton was convinced that there should be a link between print patterns, race and 'temperament'. In his book *Finger Prints* (accessible from the website www.biometricbits.com), he ponders if he might discover a 'more monkey-like pattern' in 'other particularly diverse races'. You can sense his disappointment in not finding such a link. If you do chose to read his book, be prepared to spend a lot of time angry and/or uncomfortable. I'd also highly recommend reading *Superior* by Angela Saini (Fourth Estate, 2019) for a deep dive into the history of 'race science'.

reproducibility between experts, and the risk of cognitive bias.

If you've ever watched any of those sleek 'crime scene investigation' TV shows, you might be wondering what cognitive bias has to do with anything. Surely the pattern-matching is all done by computer, right? Well, though computerised databases do play their part, it turns out that the process of comparing a print with one stored in that database is surprisingly human. Here in NZ, software is used only as an initial filter that looks at the overall pattern of the print, and at the relationship between various points on the image. That computer analysis spits out a long list of possible candidates, each of which is then examined in detail by a human – a trained fingerprint specialist. And there's plenty to look at. Tanja Van Peer, who runs NZ's National Finger Print Service, told me:

> The quantity of information in a single, good-quality latent print can be vast. When we have the image in front of us, it's not just the flow and shapes of the ridges that we look at. Pores, skin creases and scars are all unique too. Once we've narrowed down the search on screen, we pull out the original set of prints and repeat the analysis. Every identification we do also goes through semi-blind checks with two other specialists, and all of that is repeated when it goes through court. Our verification process is robust.

Yet even with all those checks and balances, fingerprint analysis is always considered **opinion evidence**. Yes, it's based on the judgement of a highly qualified expert, and the probability of them linking a print to the wrong person is very low, but it's not zero. By its nature, opinion evidence cannot offer absolute certainty. In 2017, the American Association for the Advancement of Science (AAAS) said that '[Examiners] … should avoid statements that claim or imply that the pool of possible sources is limited to a single person. Terms like "match," "identification," "individualization" and their synonyms, imply more than the science can sustain.'

However, taking the human out of fingerprint analysis is unlikely to make the process any more accurate. In fact several studies have shown that, when it comes to making comparisons between prints, well-trained examiners perform significantly better than any automated system. During my visit, Van Peer repeatedly emphasised that specialists in NZ undergo a rigorous five-year training programme to refine their skills, but she also acknowledged that even their robust approach to analysis can never be entirely error-free. A growing number of organisations are also adopting similar 'blind verification' steps to reduce the risk of bias. And there seems to be a global appetite for making the process more scientific. University of Lausanne Professor of Forensic Science Christophe Champod has argued that one way to do this is to assign numerical probabilities to fingerprint evidence, which would bring it closer in line with how DNA evidence is presented in court. Several mathematical models that aim to do this are under development, though none are ready for widespread adoption just yet.

Fingerprinting will continue to be used as a form of forensic evidence in court, but the hope is that these efforts will improve its reliability and objectivity, while also formally establishing that – like all forensic techniques – it's not infallible. The only person who should ever confidently declare 'a perfect match' between prints is a fictional TV detective.

Touch

OK, I think we've stared at our papillary ridges for long enough. Let's talk about why humans have them and what they do. Perhaps surprisingly, there's no definitive answer to this, but there are two leading theories.

We get a clue to the nature of the first one from the features' alternative name: **friction ridges**. The basis of this theory is that the complex network of raised whorls, loops and arches controls the frictional interactions between our fingers and any surface they come into contact with. In other words, they play a role in gripping and grasping. Although thousands of studies have explored this idea, their findings have been hugely varied, and often contradictory. 'Part of the reason for this is that it's hard to design experiments that will allow us to accurately measure what we need to measure, but that also reflect real-world conditions,' said Professor Matt Carré, speaking to me from his office at the University of Sheffield. Carré is a biotribologist – he's interested in the frictional interactions the human body makes with the world around it, and skin is a major focus of his work.

'Friction-wise, when I pick up my coffee cup, there's a lot going on at that interface between skin and cup. Bringing it into a lab setting involves trying to reproduce just the essentials of what's a very complex set of interactions.' For most researchers, 'the essentials' can be summed up as one finger in contact with one surface (or a set of surfaces). So over the decades, a variety of instruments have been built to provide this set-up and to measure the frictional forces at play. They largely fit into one of two groups, each one with its own strengths and weaknesses: either the surface is fixed in place and the finger moves, or the finger is stationary and the surface moves beneath it. Both approaches follow a long tradition of friction experiments that involve rubbing one solid along another while taking measurements, and many have provided some valuable insights. But for a whole host of reasons, skin is an especially challenging material to study.

For a start, just like the tyres we explored in Chapter 5, skin is viscoelastic. 'It's kind of rubbery,' said Carré. 'And the skin that's on our hands and feet is significantly thicker than it is elsewhere on the body. So the mechanics of it, and the properties that define its behaviour, are non-linear.' This means that skin doesn't exert a constant, unchanging force on objects; it deforms under load, allowing it to make intimate contact with a range of shapes and textures. Generally, the larger the load, the more skin can cling and conform to a surface, increasing the real area of contact between them. It might seem reasonable, then, to assume that the harder we press our fingers onto a surface, the higher the friction between them will be. But once we include our papillary ridges in the conversation, everything gets more complicated.

In a recent study, a group of Chinese researchers suggested that the question of whether these ridges help or hinder contact depends on how a finger is moving. They were interested in directionality, and wanted to find out if a finger sliding forwards on a surface would experience more or less friction than one sliding backwards. In their set-up, the study subject's hand was positioned so that their index finger met the surface at a known angle. A small metal bar was gently pressed into the finger from above to hold it in place. This ensured that a specific force (or load) was applied to various metal surfaces as they were moved slowly beneath the finger. In every test, a finger sliding backwards (relative to the surface) experienced a higher coefficient of friction than one that was sliding forwards.* To understand why, the researchers repeated the same experiment with glass surfaces and a finger that had been dipped in ink, and studied the resulting prints. They saw that on a finger dragging back along a surface, the papillary ridges spread out, while on a finger pushing forward, the ridges moved closer together. They concluded that it was this change in contact area – enabled by the flexibility of papillary ridge skin – that controlled the friction.

Carré, who has been studying fingerpad friction in sports and medicine for more than a decade, has also explored the relationship between skin's viscoelastic properties and its frictional behaviour. In one study, he used a suction-based device to determine the distensibility – a measure of how prone a material is to deformation – of skin at different sites across the palm. He found a link between this metric and the thickness of the outermost layer of the epidermis (the stratum corneum). The thicker the stratum corneum, the more deformable and elastic the skin. And via some force measurements, he confirmed that skin with this thicker outer layer also produced the highest friction. In a separate study, Carré showed that the

* A reminder: the coefficient of friction is a system property – it's a measure of how much force is required to make a specific material slide along another specific material. So its value is meaningless unless you know what two materials are involved.

relationship between friction and skin deformability changes depending on how firmly a finger is pressed into the surface. At low loads, the measured friction is dominated by hysteresis (see Chapter 5) – the rubbery fingerpad deforms and flows over the surface as force is applied to it. But at increased loads, the fingerpad starts to lose its rubberiness and become rigid. When that happens, friction begins to increase linearly.

Unsurprisingly, the roughness of a surface also has an impact on how well skin can make contact with it. This is why, Carré explained, 'Ridges of different designs and patterns are often added to surfaces like sports equipment and aids for people living with disabilities. The assumption is that they improve grip, but it's not clear how much science went into their development.' Working with colleagues in Sheffield and Eindhoven, Carré set out to better understand the impact of these patterns, designing brass surfaces with a range of different ridges machined into them. They measured the highest friction when the ridges were tall, narrow and widely spaced (2mm, 6mm and 10mm respectively – 0.08in, 0.2in and 0.4in) and concluded that it was due to significant deformation of the fingerpad. But the friction wasn't always high; it periodically decreased and increased as the finger travelled across the surface ridges. The most *consistent* friction was measured for smaller, more closely packed ridges. 'We found that there is no single optimal pattern,' Carré continued; 'It depends on what the item will be used for and how it is gripped. We don't always want a high friction coefficient in our daily tasks. Sometimes a lower but more reliable, consistent value works best.'

The single most significant factor in skin friction seems to be the presence of water, both in and on the surface. 'In the lab, we've seen the effects of hydration within the skin, especially on the hands,' said Carré. 'The stratum corneum – effectively a layer of dead skin cells – can have naturally very different hydration levels, depending on the individual, or even at different points throughout the day. Generally, more hydrated skin tends to be softer and produce higher friction on surfaces than dry skin.' Multiple studies support Carré's experience, and have shown that skin's structural properties are very sensitive to moisture. The material swells and becomes more pliable and rubbery as the humidity of the air around it increases. This drives up the area of contact that skin can make with an object, increasing its friction.

Sweat too can have an effect on the contact mechanics of a fingerpad. Secreted from pores along each papillary ridge, sweat can interfere with the interaction, causing the coefficient of friction between a sliding finger and a smooth surface to *increase* by about an order of magnitude. But if the surface is porous and rough like paper, friction tends to *decrease* over time. Researchers think that this may be due to the paper becoming 'smoother' as sweat gets absorbed into its pores (not unlike the damp sand of the Introduction). Regardless of the surfaces involved, there seems to be widespread agreement that there's an optimum level of moisture that aids in grip, and it sits somewhere between bone dry and totally saturated. Korean researchers recently showed that friction is maximised when a very thin film of water or sweat is present on a surface. Once the volume of water increases so that the papillary ridges are 'flooded' (beyond about 0.2mm/0.008in in their case), the frictional force plummets.

That particular study was based on a silicone replica of a finger rather than the real thing, so it omitted a potentially important phenomenon, one that's familiar to anyone who has ever hand-washed a mountain of dirty dishes, or luxuriated in a hot bath after a long day: the temporary transformation of fingerpads into wrinkly, prune-like blobs. In 2011, a group of US-based researchers showed that these wrinkles, which typically start to appear on fingers after about five minutes of being submerged in water, might help us to grip things when conditions are exceedingly wet.* The team's initial idea was that the wrinkles acted like tread blocks on wet-weather tyres, with their channels working to drain the water away from the contact area. When they analysed images of water-wrinkled fingers, they found a different analogy – there were clear similarities between the wrinkle patterns and the drainage network of many river basins. This mechanism, combined with skin's viscoelasticity, means that when you grip onto a wet object, much of the liquid could be squeezed out along the wrinkles, allowing your skin to make close contact with its surface. But whether or not these wrinkles actually aid in grip is still very much up for debate. One study that involved transferring small submerged objects (fishing weights and glass marbles) from one container to another concluded that wrinkling *does improve* handling, allowing participants to complete the task more quickly. A similar study carried out by a different research group a year later concluded that the presence of wrinkles granted *no improvement* in dexterity. And a paper published in 2016 determined that wet-wrinkles actually *reduce* finger friction and grip performance.

* This wrinkling tends to happen much more quickly in freshwater than in seawater.

So at this stage, we can't really draw any firm conclusions on the usefulness of pruney fingers. But we do know how they form, and contrary to what you might expect, it has nothing to do with water passing into the skin and plumping it up. Instead, these wrinkles are caused by the constriction of blood vessels buried deep within the epidermis. In a similar way to how a tent will collapse when its internal supports are removed, the shrinking vessels cause overlying structures to fold inwards. The complex internal structure of papillary ridge skin limits this collapse, so that it produces the characteristic deep channels that we associate with water-immersion. The leading theory for why wet-wrinkles only occur on the hands and feet has to do with the presence of sweat glands, which, below the surface, make intimate contact with a dense network of nerve endings. Chemical changes in the sweat glands, which result from being submerged in water, trigger these neurons and cause the blood vessels to constrict. This process happens without any conscious thought on our part. It's all down to the autonomic response of the nerves that fill our fingers. The link between neurons and the formation of water-induced wrinkles is so strong that for many decades, their presence has been used as a simple test of nerve function within the hand. Surgeons realised that after immersion in water, fingers that had – through injury or surgery – experienced sensory loss tended to stay smooth, while those around them wrinkled as expected. In cases where the nerve damage was temporary, once sensation began to return to those fingers, the wrinkling did too.

That discovery brings us on to the second theory for why we have ridged fingers: their patterns enhance our ability to feel tiny features on surfaces. It's time to talk about the sensors behind our sophisticated sense of touch.

Sense

Close your eyes. I'd like you to undertake a careful tactile exploration of the book (or device) that you're currently holding, and describe it by feel. How heavy is it? Is it at a similar temperature to your skin? If you press down on it, does it 'give'? Can you discern any textures on its surface? What about its edges – are they sharp or rounded? You're able to answer those questions and more because buried within your skin's many layers are a series of intricate sensors called **mechanoreceptors**. As their name suggests, they gather information about any mechanical force or stimulation applied to the skin, such as pressure, motion or deformation. They then send it off, via their attached neurons, to the central nervous system for processing.* For tactile interactions, there are four main types of mechanoreceptor, each with their own oddly poetic name.

Unknown to us, when we type on a keyboard or feel for the edges of an object, we're gathering information via our **Merkel discs**. These are bundles of nerve endings that are located fairly close to the skin's outer surface. Merkel discs are found all over the body in both glabrous and hairy skin,

* Mechanoreceptors are just one of five types of receptors that form the human body's somatosensory system, which is distributed across our skin. The other four are thermoreceptors (which detect changes in temperature), proprioceptors (which detect body and limb position and self-movement), chemoreceptors (which detect chemical changes – they help us to taste foods and drinks) and pain receptors.

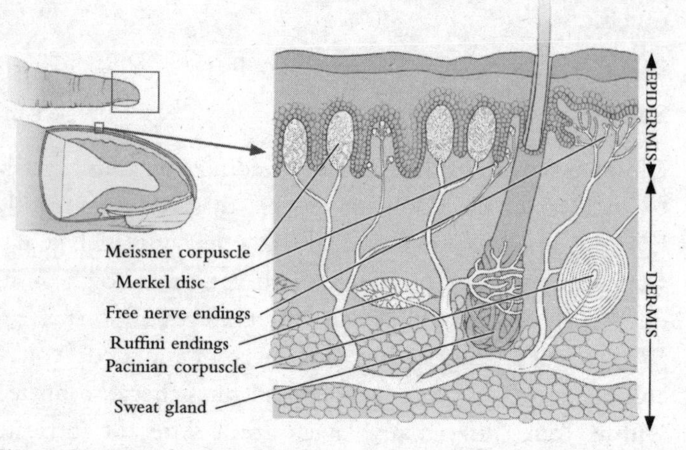

Figure 21: Your skin's remarkable touch-sensitivity comes from its mechanoreceptors.

but because they're particularly densely packed underneath papillary ridges, they're very common in fingerpads. As well as helping us to identify edges and shapes, Merkel discs respond to changing textures, and they are extremely sensitive to the tiniest of indentations. They can also sense low-frequency vibrations. Merkel discs are described as 'slowly adapting', which means that for as long as you're touching an object, they continue sensing it and transmitting that information to the brain.

This is different from **Meissner corpuscles**, which, if nothing changes, just stop sensing. These are the most common receptors found in the skin of our hands, and they're buried within the 'inward facing fingerprints' of the dermis. Though they're not particularly helpful in static touch, or when your skin is in sustained contact with an object, Meissner corpuscles are absolutely critical in dynamic touch. So any time we move our fingers along a surface, these bean-shaped corpuscles gather important tactile information. And if you've ever instinctively adjusted your grip on an object because you felt it begin to slip, it's the 'rapidly adapting' Meissner corpuscles you have to thank. They stand on their ends, perpendicular to the surface, so when the skin is compressed, it deforms the corpuscles. Some researchers believe that the shape of papillary ridges magnifies this effect, contributing to the corpuscles' sensitivity. And multiple studies have shown that people with long-term blindness tend to have especially finely tuned Meissner corpuscles. This is usually attributed to the importance of these receptors in the reading of braille.

Buried even deeper in the skin are two more classes of mechanoreceptors, both of which are in the business of sensing pressure. The first are **Pacinian corpuscles**, which, at a glance, look similar to Meissner corpuscles, but they're much larger, 3–4mm (0.12–0.15in) in diameter. They adapt even more rapidly too, and respond to even smaller deformations. This suggests that they help us to discriminate very fine textures, especially with moving fingers. Back in 2009, French researchers designed a tactile sensor with ridges that mimicked the size, shape and orientation of the papillary ridges on a human finger. When they slid it along a surface, it produced vibrations in the range of 200–300Hz, exactly where Pacinian corpuscles are at their most sensitive. This study is the source of the theory that the ridges on our fingerpads aid our receptors in their search for information. At the time of writing, it had not yet been definitively proven, though later studies have offered some support.

Ruffini endings are the final piece of the puzzle. They are almond-shaped receptors that detect sustained pressure and deep touch, as well as skin stretch and deformation in joints. They provide useful information on finger position, which makes them very important in grip control. Ruffini endings are also believed to detect warmth, which may explain why the pain of a burn seems to linger below the skin surface.

Together, this collection of mechanoreceptors allows us to instinctively manoeuvre through our tactile world.* Every time our fingers touch an object, these tiny biological machines bend and flex, and countless nerve endings fire, sending distinct electrical signals to our brain. Our palms, too, hum with information, and the position and movement of our digits add critical exploratory clues. So when we push down on an object we learn something about its hardness, and we judge its weight by lifting it. Enclosing it in our hand gives us an idea of overall shape and volume, while tracing its edges with our fingers provides more exact details. When we move our fingers back and forth we expose fine surface textures, but if we want to gauge an object's temperature, static contact works best.

Within a few hundredths of a second, all of these signals have been blended into a seamless stream of data and sent to our brain. In about as much time as it takes to blink, we've responded, adjusting our fingers and their robust but pliant skin, to maintain an appropriate level of grip. Humans react more quickly to tactile stimuli than to auditory or visual ones, which suggests that our ability to *feel* has played a vital role in our evolution, as well as in our daily lives. And for the more than 35 million people on planet Earth who live with visual impairment, touch is more than a single sense.

---

* Glabrous skin also contains free nerve endings, which are sensitive to pain, extremes of hot and cold, and light touch. A fifth kind of mechanoreceptors, called Krause end bulbs, are also believed to sense cold temperatures, but I could find very few studies into their specific function.

Dots

Sile O'Modhrain was in boarding school in Ireland when BBC Radio 4 first broadcast its dramatisation of *The Lord of the Rings*. She listened, enthralled, to every episode, but it wasn't the epic tale or even the actors' performances that drew her in. 'The sound effects were staggering – they made the series special. That was the start, really. I became hooked on everything coming out of the BBC Radiophonic Workshop, and knew that was what I wanted to do.' After a degree in music and a masters in music technology, O'Modhrain was offered the job she had dreamed of – a studio engineer, making radio programmes for the BBC.

At the time, audio recording was still done on tape, so editing involved physically cutting and then splicing sections of tape together. Just a few years later, in the mid-1990s, that all changed. 'Audio started to go digital, and so editing suddenly required sight,' said O'Modhrain, chatting to me over the phone. 'Previously, when I was editing, I could put marks on the tapes that I could feel. But now engineers needed to be able to move a cursor around on a screen to highlight sections of sound waves. For me as a blind person, this shift was huge – it effectively meant I could no longer do my job.'

O'Modhrain moved to California to do a PhD at Stanford's Center for Computer Research in Music and Acoustics (CCRMA), and found herself surrounded by 'a whole community of amazing people who were thinking very deeply about the future of digital sound'. That set O'Modhrain on a new path – one she's still on – to make music and digital interfaces more tactile. In her career to date, she's investigated everything from tangible sound maps and quantifying the 'feel' of musical instruments, to gesture control for virtual reality and web-browsing tools for blind users. And now as a Professor in Performing Arts Technology at the University of Michigan, she's working to develop a full-page braille display that can be used for both text and tactile images.

Braille, the touch-based system of reading and writing, was invented by French teenager Louis Braille when he was studying at the Royal Institute for Blind Youth in Paris. He'd heard about a code called 'night writing' that had been used by Napoleon's army. Based on raised patterns of dots arranged into large cells of 12, it allowed troops to safely (if slowly) share simple messages after dark. Louis's version, which he went on to publish in 1829, was both simpler and more sophisticated than the military code. He used six-dot cells (two columns of three dots) with different combinations of dots representing individual letters, numbers or punctuation marks. The cells were small enough that each one was encompassed within a single fingerpad, which made it quicker to read.* The basics of braille remain largely unchanged, though it has been expanded to include symbols and mathematical operators, among other things. Most experienced readers today use contracted braille, a form of shorthand where some cells represent common words (e.g. you, that, from) rather than individual letters.

A full page of printed braille typically has about 1,000 six-dot cells spread over 25 lines. Readers move their hands along each line from left to right, encountering cells sequentially, though only the index fingers actually touch the dots. There have been some studies in which people were asked to read using their index and middle fingers

---

* Braille also used this same system to develop a musical notation, with each pattern representing the pitch and rhythm of a note. The Royal National Institute of Blind People (RNIB) has a guide on braille music on its website.

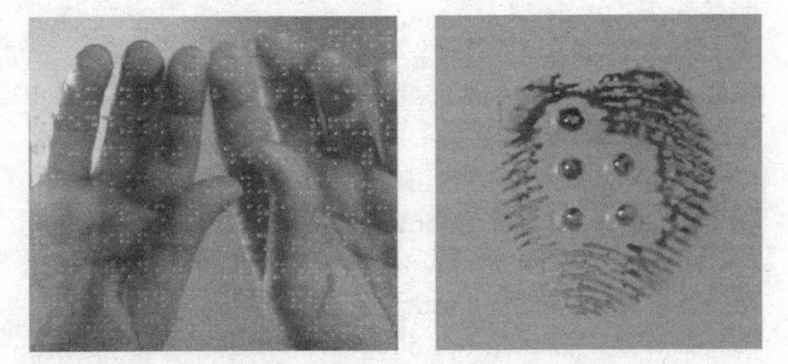

Figure 22: In reading braille, finger pads deform, taking on a characteristic shape for each letter or word.

simultaneously, but rather than making it easier, it *reduced* reading performance, and made it harder to recognise individual characters.

Braille is effectively a texture made from regularly spaced dome-like dots, approximately 0.5mm (0.02in) high. So as the fingerpad moves across a cell, the skin deforms in response to the dots, which creates a frictional force. But as O'Modhrain showed in a 2015 paper, with the light touch and lateral motion typical of braille reading, the contact patch between the fingerpad and surface changes continuously and rapidly, and the skin is not so fluid as to be able to 'flow' between the dots on a given cell. The patch, therefore, takes on 'a characteristic shape for the particular braille character being read'. There have been several studies into the relationship between this unusual contact and the response of skin's mechanoreceptors. One from a group of Swedish researchers suggested that imperceptible slips of the papillary ridges over textured surfaces could be at least partly responsible for the triggering of Merkel discs and Meissner corpuscles.

'Reading braille is a dynamic process,' said O'Modhrain. 'Psychologist David Katz did a lot of the seminal work around fingertip contact on surfaces back in the 1920s. One of the things he said was that, if you stop moving, the tactile impression or sensation fades from view. In other words, you have to move in order to feel.'* This idea, now known as active touch, says that while your mechanoreceptors can gather some valuable information from static or passive contact, the additional movement of the hands or fingers is what makes our sense of touch so exquisitely precise. When I asked if this marked the difference between sensing and feeling, O'Modhrain responded, 'Yes, in effect. When you're controlling the movement yourself, your impression of the world sharpens. The perception you get when you're actively exploring a surface is much more vivid than it would be if, say, somebody moves something against your finger.' A study to determine the role of touch sensors in printed braille reading found that active touch – where the participants could move their finger along a line of braille – yielded superior performance to any passive or static touch conditions. O'Modhrain and her Michigan colleagues came to similar conclusions after their study with a group of experienced braille readers using a single-line braille display. Identification errors were at their lowest when there was sliding contact between the fingertip and the braille surface.

This, said O'Modhrain, is why technologies that involve single braille cells refreshing beneath a stationary finger have proven to be so unpopular with braille readers. 'When you move your fingertip, you get way more information than if pin-based braille patterns push up into your stationary finger.

---

* David Katz's work on touch was published in a book called *Der Aufbau der Tastwelt* (1925). In the late 1980s, it was translated into English and published as *The World of Touch* (available as a Routledge paperback edition, 2016).

And yet sighted manufacturers continue to make such products, naively touting them as real solutions for blind computer users.' That's beginning to change. Large tech companies are increasingly, though belatedly, employing visually impaired engineers and designers to ensure that their products are (and remain) accessible. And programs driven by visually impaired researchers like O'Modhrain are spinning out of the lab and into the market. Many of these efforts form part of a larger goal, dubbed 'The Quest for the Holy Braille', which aims to produce a low-cost, full-page, refreshable, tactile display, similar to an e-reader, but with braille pins rather than pixels. 'A development like this would be about as significant to blind people as the transition from a command line to a graphical user interface was for sighted people,' she said.

A small number of single-line, half- and full-page braille displays are already commercially available. Most rely on devices called piezo bimorphs that bend when a voltage is applied to them. This action pushes pins up through a surface covered in holes, creating each braille character. Though effective, bimorphs require a lot of back-end electronics to work, and that makes the final devices bulky and heavy. Current displays are also eye-wateringly expensive, at up to $50,000 (£36,000) for half a page of braille. What's needed is a more streamlined and much cheaper device, ideally with a price tag closer to $1,000 (£700). So instead of bimorphs, the Michigan team – O'Modhrain and mechanical engineers Professor Brent Gillespie and Dr Alexander Russomanno – are using an approach based on bubbles. 'A few years ago, Brent mentioned that he'd found a way to make tiny bubbles move up and down inside a soft, flexible membrane. It made us wonder if we could use that motion to drive braille pins,' she explained. In their current prototype display, each

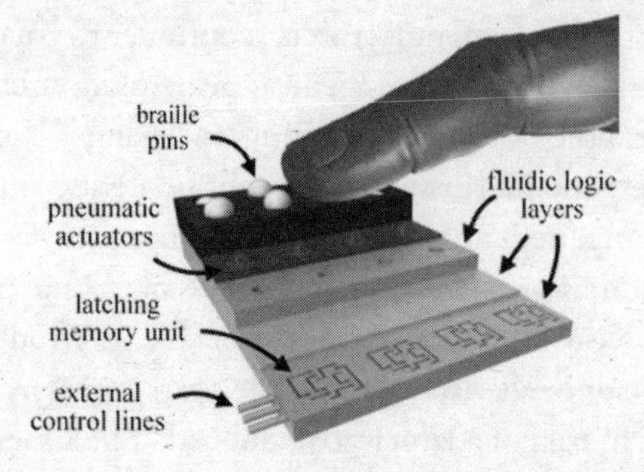

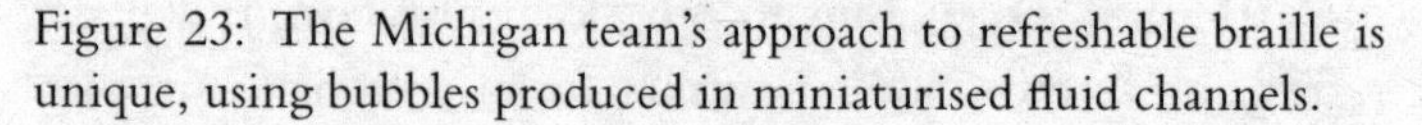
Figure 23: The Michigan team's approach to refreshable braille is unique, using bubbles produced in miniaturised fluid channels.

individual braille pin sits atop a tiny, membrane-covered cavity. These, in turn, sit on a network of microchannels that pressurised fluid flows through. When a specific braille character needs to be displayed, control valves (analogous to electric transistors) open, allowing fluid to flow into only the relevant cavities, which pushes up the pins that form the character.

'We see several benefits to this approach,' said O'Modhrain. 'For a start, despite the channels being so small that you need a microscope to see them, they're relatively straightforward to manufacture. It's mostly produced in a block, rather than assembled from individual components. It also has no mechanical parts, and the actuator (the bubble cavity) sits directly below the pin, so it takes up a tiny fraction of the space that bimorphs do.' When we chatted, the team had produced a small-scale device – about half the size of a smartphone – that could display around 200 pins. The aim is to get to at least 6,000, and O'Modhrain was optimistic: 'Everything works. We have the substrate, the logic and the bubbles integrated, and they're driving pins all under the control of the air valves. We're sorting out our supply chain too. We're pretty much at the stage where we can begin to scale up … though that will come with its own challenges!'

One of those challenges will be around choosing the pin density – the optimal number of pins-per-square-centimetre – to display both literary braille and tactile images. There's a trade-off to be made. The dot density of standard braille isn't high enough to let you render the smooth curves or diagonal lines needed for maps and charts, but if the dots in a braille cell are too close to each other, they're much harder to differentiate and therefore read. Some device manufacturers have suggested that using thinner pointy pins might be one way to increase the pin density, so O'Modhrain and her colleagues set out to see what effect this might have on a reader's perception of braille. They 3D-printed a series of braille surfaces that used six-dot cells for each character. They had three categories – normal, where each dot was represented by a single pin (either standard diameter or pointy); group, where four pointy pins were used to represent each dot; and blob, where only the overall shape of braille characters could be discerned. The blobs were unpopular with the participants, as were characters rendered with groups of pins. 'But ultimately what we found is that readers are primarily interested in the number of dots and their relative position to each other,' she said. 'So as long as we maintain those things, we might be able to make the pins thinner, and push up the density. Pointy pins aren't that comfortable if you're reading large amounts of text, but they could be good for high-density graphics.'

At the time of writing, the Michigan system used standard pins, but O'Modhrain told me that as they continue to work on the system, these higher-density, thinner pins may become the default option. Her closing statement was a positive one.

I'm hopeful that a solution for a full-page tactile display isn't too far away. Even if it doesn't come from us, it'll be someone else with another good idea, because there's a clear need for it, with millions of potential users. Speech-based technologies and single-line displays can only get us so far – they still rely on the blind person maintaining a mental model of how information is organised. A tactile display would offload some of that responsibility onto the device, and give us many more ways to interact with content. It could be transformative.

Haptic

One of the most remarkable things about the human sense of touch is that, despite its reliance on the physical deformation of tiny receptors within our skin, it's not confined to our bodies. Or at least, we have ways to extend its influence. Each time we use a tool, we get a sense of what's happening at the end of it. We can feel the roughness of paper through the nib of a pen, or detect how loose or densely packed the soil is via the handle of a spade. This ability to feel at a distance, to capture tactile clues from objects that we're not directly touching, has long fascinated researchers. It wasn't until the 1960s that it was first definitively linked to Pacinian corpuscles. Around 2,500 of these mechanoreceptors are found in each human hand, with roughly twice as many in the fingers as in the palm.

Their signature is an extreme sensitivity to high-frequency vibrations, which is why they're particularly important to this so-called 'tool-mediated touch'. Countless studies have now confirmed that as the pen nib or the spade edge digs into a surface, it produces vibrations that then transmit through the tool and into the hand's network of

Pacinian corpuscles. There are still some questions about what happens beyond that, in the central nervous system itself, but a growing body of evidence suggests that the brain uses very similar mechanisms to process these signals, regardless of whether they were generated by direct contact or mediated by a tool.

While our sense of touch is somewhat dulled by the use of a tool, we can still gather very useful information, such as how hard or stiff an object is, as well as gaining a detailed understanding of its texture and frictional properties. And just like with our fingerpads, tool-mediated touch is most accurate when it's active rather than passive – moving a tool across a surface tells us much, much more about that surface than keeping the tool still.

Heather Culbertson, an Assistant Professor of Computer Science at the University of Southern California, uses this to her advantage. Her system can scan and capture some of the essential tactile characteristics of surfaces, and reproduce them elsewhere, using a hand-held tool that's rather similar to a chunky ballpoint pen.* This technique is called **haptography**, a portmanteau of 'haptic', from the Greek word meaning 'to touch', and 'photography'. And Culbertson is one of the leading figures in this novel field. 'In essence, haptography involves taking a snapshot of an interesting haptic interaction that you would want to feel later,' she told me. 'Maybe you want to recreate the feeling of a specific fabric, or a piece of wood, clay or plastic.' Culbertson and her colleagues have even used it to record a sample of dinosaur skin from a museum collection, so that

---

* Culbertson's former advisor and long-time collaborator Professor Katherine J. Kuchenbecker submitted the original patent for the haptography pen in 2011.

schoolchildren in other parts of the country could feel it for themselves.*

The basis of haptography is accurately capturing the vibrations produced when you drag a tool along a surface – the same vibrations that excite our Pacinian corpuscles. But other signals are captured too, as Culbertson explained:

> Touch is a mechanical sense. The act of touching an object generates forces, vibration and friction. In our skin, those signals are detected by mechanoreceptors, but in our haptography system, we have a pen with a bunch of sensors inside it. As you move it across whatever object you're interested in, force sensors measure how hard you're pressing, position sensors detect where you are on that surface, and specialist accelerometers capture the distinctive high-frequency vibrations that result from surface texture and hardness.

Unlike in many touch-related experiments, the person holding the haptography pen is free to move it in any way they wish – they're not restricted to certain speeds or forces. While this makes the data processing more challenging, Culbertson's research suggests that it adds to the realism of the virtual surfaces they go on to create. 'We cannot say that an object or a surface always feels a certain way. We can only say that it feels this way if the person interacts with it like this. People have their own internal model of perception; they instinctively know that the feeling of a surface will change depending on how

---

* A talk that Assistant Professor Culbertson delivered in 2020 is available on YouTube. Search for "Haptics For Communication in a Socially Distanced World" to find it.

they move. We don't try to fight that.' Once the tactile data is captured, Culbertson slices it up into tiny segments of time – tens of milliseconds in length – and uses it to create a mathematical model of those interactions. That model can then be played back though a similar pen, so that it vibrates as it is moved along another surface. It also adjusts to match the motion of each individual, slowing down if they do, or increasing in amplitude as the person presses harder. The resulting tactile illusion can transform a display screen into a surface that feels like a brick wall, a piece of carpet or a cork board.

Culbertson has used her haptography system to capture the tactile properties of more than a hundred different surfaces, and when we spoke, she said that the resulting database had been downloaded 'several thousand times'. There are no current plans to commercialise the system, but the hope is that by making the models and the texture database free to access, the technology will support lots of different applications.

Over the past few years, Culbertson has been working on adding haptic interfaces into virtual and augmented-reality environments. In one project, her team designed a device that, when worn on the fingers, simulates the feeling of weight. This grasper, which they've (brilliantly) called Grabity, uses skin deformation and opposed forces to make the experience of picking up a virtual object more realistic. In another, they've focused on dentistry, where touch plays a critical role. 'Dentists pick up a lot of clues from touch interactions during examinations and surgery, so it's a big part of their training,' Culbertson said. 'However, dental students mostly practise on plastic models of teeth, which have fairly limited realism. We want to augment those physical models by overlaying virtual haptic signals onto them, to recreate the feeling of decayed teeth or plaque or cavities.' Haptography has also been suggested as a way to add 'touch feedback' to robot-assisted surgery, where a human surgeon's hands manipulate precision robotic instruments rather than traditional surgical tools.

A growing area of focus for Culbertson is in generating multimodal textures, which don't solely rely on haptic signals. 'Your senses don't exist in a vacuum; your perception of the world is rarely, if ever, defined by any one sense. So when you interact with an object with a patterned surface, what you can see will influence what you feel.' She recently found evidence that backs up the experience of those of us who have dragged chalk across a blackboard: when it comes to tool-mediated touch, sound is particularly important. 'Regardless of what texture was being evaluated, sound significantly increases the realism. It more completely captures the texture's roughness and hardness than haptic cues alone,' she said. 'Of course, that makes our job harder, but then that reflects the reality of the human sense of touch – it is immensely complex and so, so interesting.'

★★★

Our skin is an interface that both connects us to and separates us from our surroundings. Nowhere is that more apparent than in our hands. From the fingerprints we leave behind, to the huge range of clues we can gather from different surfaces and objects, our sense of touch truly is a navigation tool unlike any other.

CHAPTER NINE

Close Contact

Before he retired, my dad, Jackie, was a precision engineer, a toolmaker by trade. That meant that in our garage, we always had drawers full of interesting contraptions. Sliders and angles, screws and threads, levels and probes. To a curious kid like me, those toolboxes seemed like treasure chests filled with jewels. I loved picking through them and inventing uses for each piece. One evening, a family friend dropped off a heavy polished-wood briefcase to the house. 'I borrowed that from Jackie,' he told my mum, Rosemary. 'I promise I took good care of it. All the pieces are still there!' Intrigued, I peeked inside as soon as I got the chance. More than 50 rectangular blocks of metal – steel, I was to learn later – lay inside, each one nestled in its own slot, labelled and arranged by size. As I moved the case, the sides of each block caught my eye, polished so finely that they reflected like mirrors. I desperately wanted to lift one out and have a closer look. But I got nervous, so instead, I closed the lid and went to bed. The next morning, Dad took the case back to work, but over dinner he told me they were called gauge blocks and that they're used to measure lengths very precisely.

Fast-forward almost 30 years, and I found myself at a lab bench at New Zealand's Measurement Standards Laboratory (MSL), surrounded by several stacks of familiar cases. 'Wow, you weren't kidding when you said you had "many" sets of gauge blocks,' I chuckled to Lenice Evergreen, a research engineer and expert on length standards. MSL is a metrology institute, which means that it focuses on the science of measurement. Having spent my formative years as a scientist at the UK equivalent (the National Physical Laboratory), I love visiting MSL – it feels like home. And this time, I was there to *finally* get my hands on some shiny gauge blocks. 'We need to have all these sets to ensure that we can match anything that might be sent in for calibration,' Evergreen replied, as she opened several cases to show me what they contained. 'Gauge blocks are always made from robust materials, usually one of three. Hardened steel is popular in workshops and is relatively inexpensive, ceramics – that's these white blocks – they're corrosion resistant, or tungsten carbide, which shows very good wear resistance.'

Invented by Swedish engineer Carl Edvard Johansson at the end of the nineteenth century, gauge blocks (nicknamed 'Jo blocks' in his honour) are both mundane and remarkable. Evergreen told me that they're somewhat of an industrial measurement workhorse, 'the default way that most manufacturing workshops will check and compare lengths of machined components, and keep their own equipment calibrated'. They're often greasy with oil or petroleum jelly, perfectly suited to their surroundings. And you can get a basic set for a few hundred dollars. But gauge blocks are also a precision bit of kit; their two polished sides are parallel to one another and ultra-flat, with the distance between them known to a high degree of accuracy. So each one is its own measurement tool. They can also be combined in stacks to produce a huge range of different lengths, and it's then that you really get to see what makes them so special.

The process of stacking gauge blocks together is known as **wringing**. You start by choosing the smallest number of blocks that will get you to your target length – e.g. to reach 9.6mm, you might choose a 6.5mm block and a 3.1mm one –

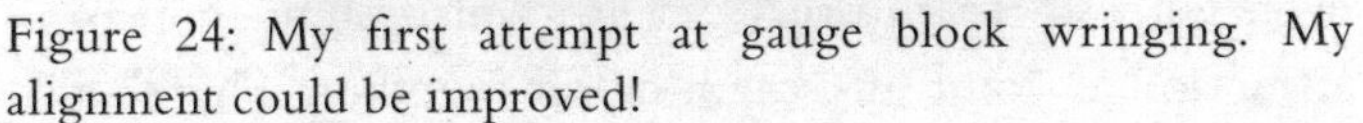
Figure 24: My first attempt at gauge block wringing. My alignment could be improved!

and clean them to remove their gloopy protective layer. 'Let's do a dry wringing first,' said Evergreen, as metrology apprentice Nina Wronski handed me a bottle of ethanol and some wipes. Evergreen then brought two newly cleaned gauge blocks together so that the polished faces met in the shape of a cruciform. Sliding them from side to side, and rotating them slowly, she gradually brought the blocks into perfect alignment, and let go of one. They stayed stuck together. 'Now you try,' she said, gesturing to the case in front of me. 'Just make sure you apply some pressure to the blocks as you do it.' I picked out two, cleaned them carefully with the ethanol and followed her instructions. My alignment left a lot to be desired, but the blocks wrung together regardless. What amazed me was just *how stuck* to each other they were. I knew to expect it, having read that during demonstrations, Johansson would suspend more than 90kg (200lb) from a pair of wrung blocks.* But nothing quite compares to feeling it for yourself. I tugged and pulled on those blocks until I was at risk of straining a muscle (what can I say, I am competitive!), but they did not budge. Wronski laughed and said, 'You need to twist them back to the cross position.' It took a bit of effort, but it worked, and the blocks came apart. She continued, 'One time when I was working with gauge blocks, I managed to wring them so tightly that I couldn't separate them even by twisting. We had to leave them soaking in a solvent overnight!'

Contrary to popular belief, the bond between well-wrung gauge blocks has nothing to do with magnetism. They're often made from non-magnetic materials, and even with steel you don't get any sense of an increasing attraction as you bring them closer together. There's also no evidence of air being pushed out forming a vacuum between the blocks, and no suggestion that a chemical reaction occurs between the materials. So what's happening?

There are two potential mechanisms that are most often credited. One is a cohesive force resulting from any fluid that might be present on the block surfaces. In the case of dry-wringing, this would be water vapour from the air, but in most practical applications, gauge blocks are wrung with some sort of intermediate oil, as Evergreen demonstrated. 'You put a tiny blip of oil on one of the blocks, and then bring them together. Now you have to slide them against each other to disperse the oil film evenly.' She handed the

---

* *The History of Gauge Blocks*, a short book by precision instrument-makers Mitutoyo, was published online in 2015 and is still freely available from the company's website.

blocks to me. 'If you slide them around now, you should feel it has the consistency of a knife through cheese?' It was a good description. 'So now we separate them and gently wipe the oil off one block, and bring them together again. Repeat that until they just stick – that's how you get a really clean wring.' Even though the volume of oil you're left with is absolutely tiny, the cohesion of its molecules produces a force that is sufficiently large to hold the blocks together. The other suggested source is some sort of molecular attraction, potentially the same van der Waals forces that the geckos of Chapter 2 exploit. This would only be possible when the blocks are ultra-smooth and completely dry, with perfect contact between their surfaces. Molecular forces may also explain why gauge blocks can be harder to separate the longer they're left in contact. 'That's why it's quite important to take your stack apart when you've finished using it, and apply a protective fluid to the blocks,' said Wronski.

Ultimately though, we can't say for sure how these blocks wring together, because we don't know. Or as metrologists from the National Institute of Standards and Technology put it, 'On a practical level we can describe the length properties of wringing films but [we] lack a deeper understanding of the physics involved in the process ... There may never be a definitive physical description for gauge block wringing.'

At this stage, you may not be shocked to hear that a mystery lies at the heart of something as common and industrially relevant as precision length measurement. After all, throughout this book we've talked about countless interesting interactions that happen on and between surfaces. We've explored the idea that when materials are in contact with one another – be it a boat moving through water or rocks colliding deep within the Earth's crust – surprising science abounds. But there's something more fundamental to consider: when we say that two things are 'in contact', what do we actually mean?

It's likely that even dry, well-wrung gauge blocks have a nanometre-thin film of water between them – so, are they really touching? And what about rough surfaces, like the running band of a curling stone and a layer of pebbled ice – how much contact are they *really making* with each other? The nature of contact also has huge implications for friction, which, despite how often we've talked about it, is still a topic shrouded in scientific subtlety. In this, our final chapter, I want to uncover some of the mysteries of surface-on-surface contact by looking only at the small stuff: zooming in to the scale of individual atoms, to make the closest contact imaginable.

Weld

First, let's take a brief excursion to Jupiter. The *Galileo* spacecraft, launched by NASA in 1989, was humanity's first foray into the atmosphere of our Solar System's largest planet. One of *Galileo*'s main communication tools was its high-gain antenna – a wire mesh dish supported by a series of ribs, not unlike a very large umbrella. It was stowed in the closed position during preparation and launch. Eighteen months after leaving Earth, motors connected to the ribs would push them outwards, deploying the antenna, and allowing the spacecraft to stay in touch with Earth on its long journey. But on the day of deployment, there was a problem – three of the antenna's ribs failed to open, leaving it lopsided and ultimately unusable. With no way to recover and repair the spacecraft, its less powerful low-gain antenna had to be repurposed. Amazingly, despite this setback, the mission would go on to meet 70 per cent of its original scientific goals. But what's most relevant to our interests is *why* the antenna got stuck. After several studies, NASA concluded that the titanium pins that held the ribs in place had **cold welded** to their nickel-alloy sockets. The metals had fused together, and no amount of force from the antenna's motors could have separated them.

Cold welding between a pair of metals is not something that happens spontaneously. It requires a specific set of conditions: first, the metals need to be entirely bare, free of surface contamination and of the native oxide layer that forms any time a metal is exposed to air. Second, at least one of the metals needs to be under physical stress, and experience some deformation. And finally, there has to be relative motion and friction between the metals. All of these were met by *Galileo*'s antenna system. The pins – coated with a ceramic and a lubricant – were tightly pushed into their sockets. Over time, this caused the pins to deform, cracking these outer coatings and exposing the metal. This was exacerbated each time the antenna was transported by truck to various NASA facilities, with relative motion between the components, and specific ribs put under a lot of stress.* During launch, as the Space

* *Galileo* spent more than four years in storage between final assembly and its launch date, which is believed to have contributed to the erosion of the coatings. This delay was a result of the tragedy that befell the Space Shuttle programme in 1986 – the loss of *Challenger* and the lives of its crew.

Shuttle punched through the atmosphere and into the vacuum of space, the pin-and-socket assembly experienced significant vibrations, causing them to rub together – but this time, in the absence of any protective oxide. To paraphrase physicist Richard Feynman, in a situation like this, the atoms 'don't know' they're from different pieces of metal, so they mix freely, and a cold weld forms.

This is contact of the most intimate kind – the two pieces of metal become one, with no obvious join or interface between them. The phenomenon was first formally described in the 1940s, by researchers investigating interactions between metals in a vacuum chamber. Samples of nickel, iron and platinum were all found to experience significant increases in their coefficients of friction as they were slid together. This was followed by 'complete seizure', when the strength of the contact matched the bulk strength of the metal; in other words, the join was mechanically indistinguishable from the rest of the material. A later study showed the same process in action in pairs of dissimilar metals, as well as in 'freshly cut titanium' in a variety of controlled atmospheres. In 1991, Harvard scientists demonstrated cold welding in the open air, between two thin layers of gold supported on sheets of a flexible solid.

Since then, much of the research into cold welding has focused on the nanoscale, exploring small numbers of atoms up close to better understand how these welds form. The motivation for such studies is not just scientific curiosity. Much of modern technology – every smartphone, inkjet printer, game controller, car, aircraft, utility meter, and a growing number of medical devices – already relies on vanishingly small micromechanical systems that involve materials moving relative to one another. A generation of even smaller devices may be on the way, and with it will come a new set of manufacturing challenges.* You can't make a structure by bolting carbon nanotubes together, or adding glue to a pile of silver nanoparticles. Instead, you use lasers and high voltages to create joins. But while they're effective, these methods are also imprecise, and the heat they produce can damage the materials. Harnessing a technique like cold welding that can be carried out at much lower temperatures could transform so-called 'bottom-up assembly' of nanomechanical devices.

In 2010, a group led by Professor Jun Lou from Rice University showed that it could be done. Their tools of choice were a series of ultrathin gold nanowires, all less than 10 nanometres (0.00001mm) in diameter. In their experiments, they placed nanowires onto two sample probes inside an imaging system called a transmission electron microscope (TEM). These probes were carefully moved towards one another and aligned so that the gold nanowires would meet 'head-to-head' or 'side-to-side'. Within 1.5 seconds of making contact, the gold in each wire could be seen to fuse together, and after 34 seconds, the process was complete – the weld looked identical to the rest of the wire, with no visible defects. To test the strength of this weld, the two probes were moved apart. Rather than breaking at the join, the gold nanowire snapped at a different location, leading the authors to

* Purely electronic devices – with no moving parts – are already made at this scale. As of 2020, Samsung were producing computer chips with components measuring 5nm across. You could fit 100,000 of them on the tip of a standard ballpoint pen (diameter = 0.5mm).

conclude that 'the as-welded structure is as strong as the original nanowire'. Lou and his team also tested the electrical properties of their welded wires by passing a current through them. There was no difference between the conductivity of the original nanowires and those that had been cold welded together. The two metal structures really had become one.

Lou credited a couple of mechanisms that are now accepted as the basis for cold welding. One is **diffusion**, which involves the gradual migration of atoms between two adjacent materials. In solid metals, this can happen at room temperature. The thermal energy that keeps atoms jiggling at all temperatures above absolute zero can also enable them to move. In the case of gold-on-gold contact, those atoms can even bridge tiny (< 0.3nm) gaps. It's this process that gives the weld its initial stickiness. The next step is **surface relaxation**, where the rest of the nearby gold atoms – which in the nanowires are arranged in a grid called a lattice – readjust their position in order to accommodate the new atomic interlopers. This process is slower, but it's what solidifies the weld, and makes its structure match the rest of the nanowire. Lou found that two nanowires with similar lattice orientations welded faster and more easily than those with mismatched lattices. And results were similarly promising for pairs of silver nanowires, as well as gold and silver combinations. This study was the first to experimentally demonstrate cold welding on the nanoscale, but it certainly wasn't the last. A flurry of others were to follow, each exploring a different aspect of the process. Some used cold-welded silver nanowires to build flexible, transparent electronic devices. Others have designed the nanoscale equivalent of sleeve bearings – components that are frequently used to reduce friction between sliding objects. We're still a little way from fully harnessing this room-temperature, atom-by-atom welding in commercial devices, but progress is relatively swift.

Lou's study did more than open the door to a new tool for nanotechnology. It gave us a unique view into what can happen when two materials meet. Through the videos the team produced to accompany their paper, we could see atomic interactions as they occurred, and watch them evolve over time. But even a technique as impressive as TEM has its limitations. Its images are two-dimensional; a cross-section of an interface, not its overall shape. If we want to study atomic interactions and probe the nature of friction, that slice won't offer enough information. We need to take a different approach.

Lube

Professor Ashlie Martini, a mechanical engineer from the University of California Merced, told me:

> There are still questions out there about what happens in the contact between surfaces, despite the fact that people have been manipulating it for millennia. The reason for that is pretty straightforward; you can't actually see the contact region or measure it directly. It is, after all, trapped between two solid surfaces. So we have to infer what's happening via indirect routes – detecting a change in temperature, friction, electrical conductivity, or adhesive strength. And then we put those into mathematical models that may or may not be true. Given the challenges involved, it's not really surprising that defining contact is so contentious.

This is especially true on rough surfaces, where only the high points actually touch, making the 'real' contact area between those surfaces a tiny fraction of the total area. And as we saw in Chapter 5's tyres, when we apply an additional compressive force to contacting surfaces, we can sometimes increase the real contact area, bringing the atoms in those surfaces even closer together, and ultimately increasing the friction between them.

That friction can be hugely costly. According to a recent study from two professors of tribology (which, as you might remember, is the science of rubbing or sliding), each year more than a fifth of the world's total energy consumption goes towards overcoming friction. In the transport sector alone, friction losses represent about 30 per cent of all energy use. Industrial applications tend to take a decidedly low-tech approach to managing friction, as Martini explained: 'The purpose of most mechanical systems is to facilitate motion in some way. Lubricants are the easiest way to enable that, because they reduce the coefficient of friction between surfaces in relative motion.'

Typically applied as a liquid or solid, lubricants physically separate surfaces from one another, giving them something more slippery on which to slide. It's likely that humans have been lubricating contacts for 4,000 years. Traces of animal fats have been found on the axles of Sumerian chariots, and, as we mentioned in the Introduction, paintings on the walls of ancient Egyptian tombs suggest that liquid lubricants aided in the transportation of heavy objects. Modern lubricants are considerably more varied and exotic, with specific formulations available for every apparent need, from the turning of wind turbines to the spinning disk of your computer hard drive.

'They're effective,' said Martini, 'but for almost all of history, they have been designed, developed and refined through a process of trial-and-error, rather than a scientific approach.' This is starting to change, and it's partly thanks to our ever-growing hunger for smaller and more efficient devices. 'It used to be the case that you'd use as much grease as possible to keep the contacting surfaces apart. But thick layers of lubricant are viscous and gloopy, which can drive the friction up – exactly what you don't want. Friction is high at the other end of the scale too, where there's no lubricating layer.' Somewhere in the middle, Martini told me, is a sweet spot – a lubricating film that, if you choose your materials carefully, can keep friction between surfaces low, despite separating them by no more than a few tens of nanometres. 'Thin lubricating films sit right at that boundary, so they offer remarkable energy efficiency with a small volume of lubricant. Saving energy is a key focus everywhere right now, so economic and environmental drivers are really pushing us to look at these materials again, only this time from a more scientific perspective.'

Martini is well placed to aid in this effort. For more than ten years, her tribology lab has looked at friction, lubrication and wear across all scales, in an effort to better understand their underlying mechanisms. Two of the many materials she's explored are solid lubrication superstars – molybdenum disulfide (MoS_2) and graphite, the same form of carbon that's in your pencil 'lead'. Both are layered materials, which means that they get their slipperiness via similar mechanisms. The stacked atomic sheets that make up their structure are loosely bound to one another. When you apply a shear force to these materials, their sheets slide across each other, similar to what happens when you push over a deck of playing cards, albeit with a lot less friction.

Where the materials differ is in their response to oxygen. Graphite works well in atmospheric conditions because when the edges of its sheets form bonds with oxygen in the air, they get smoother and slide more easily. The opposite is true for MoS_2 – reactions with oxygen cause its sheets to clump together and seize up. As a result, these materials need to be used in very different ways. You can find graphite lubricants in most hardware and electronics shops, whereas MoS_2 tends to be reserved for space applications.* The thickness of a lubricant coating will vary depending on the specific need, but between components in a precision-made mechanical system, something on the order of 2–5µm, or micrometres (2,000–5,000nm, or 0.002-0.005mm) would be fairly typical. Much too thin to be discerned by the naked eye, it's still enormous by atomic standards. The typical way to test a coating like this is to rotate or slide a sample of it beneath a small, hard pin, and measure how the coefficient of friction changes over time.

The nanoscale equivalent, used to poke and prod at the atomic layers present in thin-film lubricants, is called an atomic force microscope, or AFM. As its name suggests, this microscope can examine the force between atoms; more specifically, it can feel the forces that act between the atoms at the end of a very, very sharp tip (on the probe), and those on the surface being studied (the sample). We'll come back to the details of how an AFM works, but for now, picture the probe as a miniaturised arm and stylus from a record player, with a tip that is just one or two atoms wide at its apex. As the tip is scanned across a sample surface, it builds

---

* Martini collaborated with NASA's Jet Propulsion Lab to develop and test a specific type of MoS_2 dry lubricant for use on their Mars rover, *Perseverance*. To watch her give a short talk on that research, search for "Ashlie Martini dry film lubricant life" on YouTube.

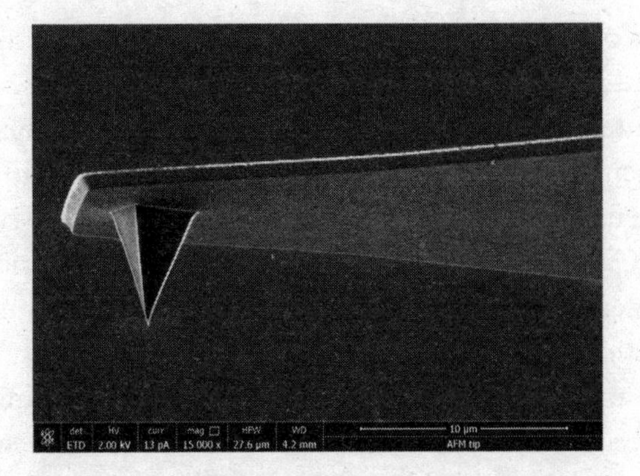

Figure 25: An image of a typical atomic force microscope tip. The bar at the bottom of the image is 10 μm or 0.01 mm in length.

up a 3D image of that surface, line by line, as well as measuring any number of different properties (depending on the set-up) including the frictional force. This tool allowed scientists to determine that both graphite and MoS_2 retain their lubrication performance, even when there are very few of their atoms present in the contact.*

If you've ever heard of a material called graphene, you might already know that it's a one-atom-thick sheet of carbon. That also makes it graphite's nano-sibling – your pencil lead is essentially made up of hundreds of thousands of sheets of graphene. But when one sheet is in isolation from others, it gains many impressive properties – optical transparency, super-strength and high conductivity to both heat and electricity. In 2011, a group of Korean scientists put

* This is not a given. Many materials change their behaviour on the nanoscale. One of the reasons gold is used in jewellery is its stability – it doesn't oxidise or react like many other metals. But nanoparticles of gold can be so active that they're used to speed up other chemical reactions.

a series of graphene films through kinetic friction testing, to find out if they could add 'good lubricant' to that list.

While surfaces coated with single sheets of graphene did experience reduced friction, thicker graphene films of two–ten sheets displayed even lower friction. Researchers from Columbia University who measured the frictional characteristics of graphene, MoS_2 and two other atomically thin materials made similar observations. In their experiment, regardless of the material, once the number of atomic sheets dropped below about five, the friction force began to increase. This received further support in 2019, when German tribologists used graphene to lubricate the contact between ball bearings and a steel surface. There, graphene flakes that were approximately three to five atomic sheets thick produced the greatest overall reduction in friction. Five sheets of MoS_2 measure just over 3nm; for graphene, it's closer to 1.7nm. We're talking about truly tiny dimensions, but even at this scale, it seems that if ultra-low friction is what you're after, having a few slippery surfaces is better than one. That bodes rather well for the potential use of these materials as lubricants in miniaturised mechanical devices. But in our own personal quest to get ever closer to understanding friction, one sheet is all we need.

Layer

Just like the pins used to do larger-scale friction measurements, AFM tips can get damaged from sliding across surfaces. They're also very sensitive to changes in surface chemistry and humidity, which makes it hard to reliably compare measurements between different samples, or even those carried out on different days. Martini took inspiration from this and designed a unique experiment – a head-to-head of the two low-friction nanomaterials, graphene and MoS_2. 'No one had ever been able to measure the frictional properties of both materials in the same AFM scan,' she said. 'But my collaborators Rob Carpick and Charlie Johnson from Pennsylvania designed an amazing sample and experiment that allowed us do just that.' The sample featured a single layer of graphene, partly overlaid by a single layer of MoS_2. In each AFM scan, the tip slid across both materials, plus the overlapping area, and measured the friction. Even though the measurement was carried out in an oxygen-free environment so as not to disadvantage MoS_2, graphene still outperformed it, providing the lowest friction at all loads for their particular tip.

This and other recent AFM studies unveiled something else. When it comes to friction, sheets of atoms aren't as smooth as you might expect. Back in Chapter 6, we talked about stick-slip friction and its role in earthquakes, but since 1987, it's been known to exist at the nanoscale, too. Researchers from IBM who were scanning an ultra-flat graphite sample with an AFM tip found that, rather than remaining uniform across the surface, the frictional force varied significantly, increasing and decreasing with a consistent zig-zag pattern. This suggested that the tip's motion was stop-start, sometimes moving with the surface, and other times sliding across it. More than that, they realised that the periodicity of this stick-slip friction – the distance between the peaks on their graph – matched the distance between the carbon atoms in the graphite lattice. This study was significant because it was a clear demonstration of friction's atomic origins. Friction didn't just happen on rough surfaces, where microscale bumps would grind along and into one another. It had deeper roots, and acted even between the flattest of flat surfaces, devoid of any roughness. These resistive forces, it seemed, were so intrinsic to materials that they could be influenced by atomic structure. As such, the results of this study challenged much of what was known about friction at this scale, and it inspired updates to the theoretical models used to describe what happens when atoms meet.

Since then, the IBM experiment has been successfully reproduced on a wide range of other materials, including MoS_2, mica, gold and sodium chloride (otherwise known as table salt), and in a variety of environments. This suggests that the effect isn't just a superpower of carbon materials. But nestled among the supportive studies are those that came to contradictory conclusions on everything from the effect of the size, stiffness and shape of the tip, to the exact mechanism by which energy is stored and released in this intermittent contact. The physics of atomic-level stick-slip friction isn't entirely nailed down yet.

A finding that drew a lot of attention was that friction measured on atomically flat materials was sensitive to the tip's sliding direction. This led people to wonder if atomic friction between moving surfaces could be 'tuned' by changing the relative positions of atoms. Dutch scientists were the first to experimentally demonstrate this effect in 2004. They rotated a graphite sample beneath a tip, measuring the frictional force at each sample orientation. When they plotted their results, they found two narrow peaks of high friction, separated by a wide region in which friction dropped to almost zero. After various tests, they concluded that the tip had somehow picked up a graphite flake, and the two high-friction orientations were due to the perfect alignment of the flake and sample's atomic lattices. In the position where those lattices were mismatched, friction all but disappeared – the graphite flake could slip over the sample with no resistance.

The analogy often used to describe this phenomenon is a pair of egg cartons sliding along one another. In some positions, the two corrugated surfaces will interlock, maximising friction and causing all motion to grind to a halt. However, if one egg carton is rotated slightly, the pattern of peaks and troughs no longer aligns. So rather than a harsh, juddering motion, the cartons can slide with relative ease, and the friction between them falls significantly. Dubbed **structural superlubricity**, this effect has been touted (mainly in breathless media releases) as the solution to all our friction problems: a means to defeat energy loss in mechanical systems.

'I'm not that enthused by superlubricity,' said Martini, when I asked her about its implications. 'I mean, it interests me because I study atomic friction, but I'm suspect of some of the hype around it. I could be wrong, but I just don't think that many of these studies are reproducible in real mechanical components under practical conditions.' Having dived headlong into the scientific literature on structural superlubricity, this seems to me to be a pretty fair criticism. Because the effect is so sensitive to the relative positions of individual atoms, it tends to occur only in highly controlled environments on pristine surfaces. The presence of any contaminants or water vapour will interrupt it, and even elevated temperatures can have an impact. In addition, the published research on superlubricity is dominated by a single class of materials – carbon, in all its various forms. A cursory glance on Google Scholar will uncover papers on everything from the interactions between diamond nanoparticles and graphene, to the sliding apart of nested carbon nanotubes.* My guess as to why carbon-based materials are so favoured? Graphite has been used as a lubricant for decades, so its properties are well known. And thanks to recent developments in materials synthesis, it's now relatively easy to make a whole host of carbon nanostructures.

But, said Martini, there's still a missing piece.

> In practice, you can never have a perfect, infinitely large layer of graphene to slide across. There will always be defects in the material, such as step edges – they're where one layer stops and another starts. We wanted to show that, look, even with a super-material with super-properties, you'll have things to overcome. The main challenge will be a sudden change in friction at these ubiquitous step edges.

In 2019, Martini and her collaborator Seong H. Kim published a series of papers looking into this subject, and the results surprised them.

Their experiment involved an atomically flat graphite surface which had a single sheet of graphene covering part of it. This provided a step, 0.34nm in height, that they could scan with an AFM tip and measure the friction force. In some scans, the tip would step up from the graphite onto the overlaid graphene. In others, the tip would step down. In terms of what they expected to happen, Martini drew parallels to how we humans go up and down stairs: 'When

---

* A Google Scholar search, carried out on 18 December 2020, found 3,250 results (excluding patents and citations) on carbon-related structural superlubricity. There were 4,140 results in total for 'structural superlubricity'.

you take a step up, to lift yourself from one step to the next, it takes energy. Going down a step is easier because gravity assists you.' Although gravity isn't at play between sheets of graphene – the mass of individual atoms is so small that the force doesn't exert any real pull on them – she explained that the height change could still have an impact. 'If we looked purely at the topography of the set-up, we'd expect to measure resistance or high friction when the tip slides up a step edge, and then low or no resistance when it slides down.'

On each of the two flat surfaces they measured an exceptionally low sliding coefficient of friction: 0.003, less than a tenth of the friction that exists between two sliding pieces of famously slippery Teflon.* This confirmed that the carbon materials were superlubric.

Near the step edge, that situation changed. As the tip climbed up the step, they saw a huge spike – friction increased by more than 30 times. So far, so predictable. Things became considerably more complex when the tip stepped down. 'We found that there was resistance on the way down too! It was much smaller than on the step up, so the topography did help, but friction still increased,' said Martini. There was some other effect at play, and it was counteracting the boost the tip received on its descent.

It was chemistry. These experiments were all carried out in air, which, you might remember from earlier, is where graphene performs best. The reactions that occur between the exposed carbon atoms on the edge of each sheet and the oxygen–hydrogen (–OH) groups in the air around them are what allow the sheets to slip over one another. However,

* See Hypertextbook.com for a summary table of values of μ for Teflon-on-Teflon.

those chemical bonds have a very different effect on an AFM tip. As the tip slides across the upper surface and towards the step edge, it suddenly finds itself facing a new chemical landscape: a band of –OHs clinging to the precipice. They exert an adhesive force on the tip, creating hydrogen bonds with its atoms, tugging on it and slowing it down. This was the source of the small, unexpected resistive force (the friction) that Martini and her colleagues measured during step-down experiments. The adhesive effect was even larger when the tip approached the step from the lower surface, because there the tip met the band of –OHs side-on, making it a much larger target for chemical bonding. This, combined with the topography – the act of climbing up a physical barrier – explains why the friction measured in the step-up tests was so much larger than in the step-down.

'So do these big friction spikes at step edges toll the death knell for structural superlubricity?' I asked Martini, as we reached the end of our chat. She said,

> Well, maybe not. Something we teased at in the paper is that the chemistry at these step edges can be controlled. If you find the right thing to terminate the ends of a graphene layer, it's possible to tune the friction and get rid of those spikes. That way, you wouldn't be limited by the defects that you find on every single graphene surface. I think this is an approach that could get us closer to a more practical version of superlubricity; one that works in real conditions.

This study – and others that came before and after it – highlights another factor relevant to our quest. It shows us that, rather than being a fundamental force, friction is an umbrella term. It's a way to neatly package together different interactions at different scales, so that they can be described by a single number: the coefficient of friction, μ. In reality, friction has (at least) two distinct components that work together to resist relative motion between surfaces.

One is **physical**, resulting from collisions between any roughness or texture on surfaces, and the other is **chemical**, when the atoms of sliding surfaces get so close to one another that bonds form across the interface. The former dominates the mechanics of large-scale structures, such as the friction between two tectonic plates. But thanks to tools like the AFM, we now know that the latter plays a critical role at the nanoscale, when the moving parts in question are just a few atoms in size. As we aim to build tiny, energy-efficient devices and components, this aspect of friction will become increasingly important. And unless we can understand exactly where it comes from, we won't be able to control it.

Contact

For 35 years, the AFM has allowed us to probe materials in unprecedented detail* – to produce 3D images of atomically flat surfaces, to tug and pull at individual atoms and to help us better understand what happens between materials. To take an AFM measurement, the probe – which, you'll

* The precursor of the AFM, called the Scanning Tunnelling Microscope (STM), was invented in 1981 by two IBM scientists, Gerd Binnig and Heinrich Rohrer. The two men went on to share one half of the 1986 Nobel Prize in Physics, with the awarding committee describing their invention as 'completely new' and saying that it would open up 'entirely new fields ... for the study of the structure of matter'. That same year, Binning published a paper with two other scientists, Calvin Quate and Christoph Gerber, which reported on their invention of the AFM.

remember, is a long lever with a pyramid-shaped tip on its end – is gradually lowered towards a surface. As it approaches, the tip's atoms start to feel those on the surface, via a series of forces. Van der Waals forces arise when the tip and surface are separated by about 10 nanometres, and just like for the gecko, they are attractive, which draws the tip ever closer to its target.* But as the distance to the surface shrinks further, another force kicks in: one that acts to repel the tip. The source of this repulsion is something called the Pauli exclusion principle, which (without wishing to delve into very muddy water) is related to the fact that, to an atom, all electrons are indistinguishable, and no two can occupy the same state simultaneously.† This means that once the electrons from the tip get sufficiently close to those of the surface atoms, their electrons will begin to repel one another, causing the probe to flex upwards.

According to Philip Moriarty, a Professor of Physics at the University of Nottingham, the point at which the van der Waals *pull* meets the Pauli *push* is an important one: 'That's fundamentally what we mean when we say things are

---

* For AFM measurements carried out in ambient air, capillary forces – produced by a nanometre-thin layer of water vapour present on the surface – also pull on the tip, drawing it to the surface.

† The laws of physics change – and get far less intuitive – once you get down to the scale of individual electrons. The Pauli exclusion principle is a quantum phenomenon, and given that neither Wolfgang Pauli (who formulated it) nor the much-lauded physicist Richard Feynman could explain the principle in simple terms, I don't feel too guilty about keeping my description short! It is, however, related to the more familiar, larger-scale electrostatic effect, where like charges repel and opposite charges attract one another. Philip Moriarty (quoted in the next paragraph above) has written about the Pauli Exclusion Principle on his blog. I'll put a link on my website.

"in contact". Whenever the net force is zero, so when the attractive and repulsive forces balance out, the AFM tip is officially touching the surface. If you try to push them any closer to one another, all you'll do is strengthen the repulsive force and they'll move apart.' Professor James Batteas, a chemist from Texas A&M University, gave me a similar answer when I asked him to define contact, as did University of Cambridge materials scientist Professor Rachel Oliver.

This tells us that at the atomic level, the contact between two solid materials is really defined by a balance of opposing forces. That alone is not surprising – after all, a mug can only sit on a table because the two objects exert an equal but opposite force on one another. But it gets less intuitive when we say that this 'touching' is mediated by electrons, the tiny, almost massless particles that eternally whizz around atomic nuclei. I think that's partly because we humans tend to consider real-world objects as resolutely solid. The book in your hand, the ground beneath your feet, the sandwich you munched at lunchtime, even the fast-flowing water in your local river. They're all substantial, sturdy, *real*.

It's natural then that we might view atoms in the same way: as hard, dense spheres with smooth outer shells. But in truth, atoms are mostly empty space. So much so that their structure has been compared to a fly in a vast cathedral dome, with the fly representing the nucleus and the walls its distant electron clouds.* And because electrons follow

---

* This is a paraphrase of something that Ernest Rutherford, the NZ scientist who discovered the nucleus, reportedly said. *The Fly in the Cathedral* is also the title of an excellent book on the race to split the atom, written by Irish journalist Brian Cathcart (Farrar Straus Giroux, 2004). These days, electron clouds tend to be described as 'quantum fields', but for our purposes, I'm not sure that adds to the understanding of what they are.

the weird laws of quantum mechanics, it's not possible to know both their exact position and their momentum at any one time. The best we can do is describe them in probabilities, which makes the outer boundary of an atom more fuzzy than smooth.*

Regardless, this finely balanced atomic tug-of-war really does define the closest possible contact between two solid objects – it's the moment their electron clouds begin to interact. Many internet commenters have taken this to mean that we can't ever truly touch something, using everything from pairs of footballs to grasping hands to illustrate the point. But Moriarty told me, it's a matter of how you define it.

> Every regime or scientific specialism will have its own way to describe contact; one that is appropriate to their needs. But if we're talking about atoms and the quantum world, analogies will only get us so far. We can't easily describe the contact between atoms using examples from the macroworld, because it doesn't scale. We need an agreed-upon scientific definition, and the point where the forces balance, that's something we can measure. So it's a definition – I'd argue the *only* definition – that makes sense.

Contact, it seems, is all about electricity.

This is also true of atomic friction, because when we describe a process as chemical, what we really mean is that

---

* This is described by the Heisenberg uncertainty principle, and results from the fact that electrons act as both waves and particles. For more on this, and on quantum mechanics in general, I recommend *How to Teach Quantum Physics to Your Dog*, by Chad Orzel (Oneworld Publications, 2010).

it involves electrons. When an AFM tip is dragged across an atomically flat surface, it's the electron clouds of those interacting atoms that control the formation of temporary chemical bonds between them. They provide the resistance that Professor Martini measured in her experiments. If two sliding surfaces – perhaps a pair of highly polished gauge blocks – are then held in contact for an extended period of time, those bonds can become semi-permanent, making it much harder to break them apart in future. And if those surfaces are the ends of two perfectly clean gold nanowires, they can fuse together so tightly that the join between them disappears.

This relatively neat picture of friction-powered-by-electrons omits an important phenomenon that will be familiar to anyone who has ever rubbed their hands together on a cold day: sliding friction generates heat. The mechanism behind this transformation from kinetic energy into thermal energy is arguably the most studied mechanism in tribology. The general idea is this: each time two solids slip against one another, their surface atoms move in opposing directions, which generates vibrations throughout the rest of the lattice. These vibrations, known as phonons, are usually described as 'atomic sound waves', but they're also the means by which heat flows through solids. In this context, the only real difference between these seemingly disparate forms of energy is the speed (or frequency) with which atoms vibrate. If the phonon frequency is low, then you're dealing with sound energy; if it's high, that corresponds to heat energy. In most contact situations, phonons of a range of different frequencies are generated, and they bounce around within the atomic lattice in all directions. Just like the waves on a stormy sea, phonons can interfere with each other, sometimes amplifying one another, sometimes cancelling each other out. Phonon transport is chaotic and messy, but it's very effective at transporting heat.

The first suggestion that phonons contribute to thermal energy loss in sliding contacts was made in 1929, and it caught on, featuring in many other friction models.* But it wasn't until 2007 that it was first confirmed experimentally, by Professor Robert Carpick – Martini's aforementioned collaborator – using an AFM. Nowadays, there's widespread agreement that generation of these lattice vibrations is what drives up temperatures on sliding surfaces. And according to Associate Professor Jeffrey L. Streator, phonons are so central to these interactions that their presence can alter the friction force that's measured.

The Georgia Tech tribologist recently published a study in which he simulated the contact of rigid sliders (grids of atoms) on an atomically flat slab of elastic material. He found that two identical nanoscale sliders in close proximity to one another measured *different values of friction on the same surface*. This had nothing to do with superlubricity; there was no magical alignment between one slider's atoms and the slab's. Instead, Streator told me, 'phonons are primarily responsible for this difference. The fundamental connection between friction and phonon propagation has been known for a long time. But what my results suggest is that it perhaps has some unexpected features.' It seems that as phonons are

* This was proposed by physicist Dr George Arthur Tomlinson, who is credited with developing one of the earliest – and still most important – models for dry friction between solids. Another physicist, Dr Ludwig Prandtl, published his own, very similar model a year earlier (1928). It is now called the Prandtl–Tomlinson Model in their honour.

generated by each slider, they move through the material. And because the material in Streator's simulation is elastic, its atoms act as if they're joined by springs, and bounce around to differing degrees. When we're talking about interactions at this scale, even tiny, temporary changes in atom position would be enough to shift the balance of forces that we use to define contact.

Atomic friction, then, is really the result of two mechanisms – one mediated by electrons, and the other by phonons. Together they produce a resistive force that opposes the relative motion between two sliding atomically flat surfaces, while transforming that kinetic energy into heat.* Neat, right? Well, in materials that are electrically conductive or magnetic, it's likely that additional mechanisms contribute to the interaction-usually-known-as-friction. If either one of the surfaces is contaminated, even with just a few atoms or a charged particle or two, this too can change their frictional behaviour. And if the surfaces are rough and textured rather than smooth, then wear plays a major part in energy loss – much of the sliding energy goes into physically removing material from a surface, as their features collide. The truth is that there's no simple answer to the question 'what is friction, *really*?' It exhibits its most fundamental form – that of fuzzy electrons and jiggling vibrations – at the atomic level, but those same mechanisms don't fully describe what we see at larger scales. And herein lies the problem with friction: there's a gap in what we know.

---

* For a growing number of researchers, this description doesn't go far enough. Theories of 'quantum friction' have been under development since at least the 1960s. At the time of writing, even the existence of this more fundamental manifestation of friction force remains a matter of fierce debate.

There's plenty of evidence that humans have recognised the phenomenon of friction for millennia, and that our ancient ancestors used that knowledge to skilfully control interactions between surfaces – from rubbing flints together to spark a fire, to wetting sand before attempting to slide huge stone blocks across it. Later, the Greeks, including Aristotle, became fascinated by the forces that govern motion, and proposed some early ideas on what they might be, largely skirting around the concept of friction. The first scientific description of this resistive force didn't arrive until the fifteenth century, courtesy of Leonardo da Vinci. But because he didn't publish his equations, the rest of the world had to wait for another 200 years, for French physicist Guillaume Amontons' friction laws, a rediscovery of da Vinci's. Since then, friction models have become increasingly sophisticated, providing a way to describe immensely complex interactions – like deformation, wear and lubrication – in the language of mathematics, and experimental studies have enriched those models.

At the risk of sounding overly evangelical about tribology, this knowledge is what powered the Industrial Revolution, allowing engineers to design new bearings, gears and other mechanical systems that were many times more efficient than what had gone before. Railways in particular flourished thanks to a better understanding of lubrication. And today, every system with moving parts is designed around friction, with some components working to minimise it, while others tap into its stopping power. Our knowledge of sliding surfaces has helped us to better understand the destructive power of earthquakes and the behaviour of ice. And the ability to measure what happens when solid meets fluid has given us highly specialised paints and adhesives, low-friction coatings and supersonic flight. It's even the basis of many of our favourite sports.

What all of these examples have in common is scale – they're big, at least relatively speaking. Yes, I know the brakes on your bike are a little smaller than an earthquake fault, but in terms of friction, they can largely be described by the same laws. And in practical terms, those laws are very well understood. After all, they've got us this far. In recent decades, however, we've made a major shift. Thanks to the development of tools that can probe the atomic world, we've scrutinised more fundamental aspects of friction, and unveiled the mechanisms that underlie it. We can model and measure how friction operates between atoms, and (we think) we know where it comes from. So on the nanoscale too, we have a deepening understanding of the process.

But these two domains sit on opposite sides of a chasm, with no bridge between them. The friction laws they each follow are undoubtedly effective, but they're also entirely separate. To date, there is no set of equations that can embed our newfound knowledge of atomic friction into our larger, more classical descriptions of the force. If such a model could be developed, it would be transformative, and I promise this is not just a case of scientists-always-thinking-that-their-pet-topic-is-the-most-important-topic. The implications of a unified description of friction would be felt far beyond the lab. The main thing it could offer is a way to predict the coefficient of friction, μ, for any pair of materials – something that is currently impossible.

This number, which tells you how easily two surfaces can slide along one another, is used everywhere, as we've seen throughout this book. But it has always been measured experimentally, whether it's steel-on-ice, rubber-on-asphalt or leather-on-oak. That means its value is at the mercy of other factors – the stiffness of the materials, the temperature and humidity, the surface roughness or the presence of contamination; they can all potentially alter the sliding friction force. Because of that, the values of this coefficient published in engineering textbooks are approximations; guidelines rather than precise quantities. And as we've discovered, those same factors that affect measurements on bulk materials are even more critical when your interfaces are just a few nanometres in size. Even if we could accurately quantify every single one of those individual factors, there's nowhere to put them; we don't (yet) have a model into which we can dump the numbers that'll then pop out a nice consistent value of μ. Or, as Jeffrey L. Streator put it, 'If you gave an expert in friction the material properties of each of a pair of bodies along with their overall geometry, surface topography and surface energy, she would not be able to tell you with any confidence what the coefficient of sliding friction would be, since there are no established predictive models.'

Without this ability, we'll struggle to build the tiny robots that countless movies and sci-fi books have promised us; at the moment, we can't even make a nanogear that will rotate freely before wearing itself out.* But even up here, on the scale of buses and conveyer belts, being able to predict μ would make it easier to design precision instruments. While most engineers won't ever need to worry about phonon transport, having it built into their equations will only improve performance. Long-term, it might even reduce reliance on lubricants,

---

* In March 2020, researchers from RMIT University in Australia reported 'stable rotation' of a tiny gear based on carbon nanotubes. It only works at temperatures below 100K (that's -173.15°C).

without affecting the efficiency or reliability of mechanical systems. As Martini put it to me, 'When you think about it, it's pretty surprising that everyone still uses a friction coefficient that's totally empirical and entirely non-reproducible. It'd be better for everyone if we could predict it from first principles.' But, she said, changing that situation won't be easy.

> We have to find a way to bring the understanding that we have at smaller scales up into the larger-scale models used in engineering. Honestly, this is a major challenge, but it is also an opportunity. For me, working in this area, it's cool that there are still open questions that, if we can answer them, can have huge impact on the world.

★★★

When I set out to write this book, I wanted to introduce you, my lovely reader, to a world that hides in plain sight: surfaces. Those places where one material meets another. I knew this story would include geckos and braille, swimsuits and tyres, earthquakes and sound barriers, ice and paint, and that the exact details would require a lot of research. But from day one, I had a clear vision for this final chapter. It would focus on how poorly we understand friction, explain that we really don't know what contact means, and show that if you look at material surfaces closely enough, the questions outweigh the answers. But, as I hope the preceding pages have shown, I've changed my views on all of these things.

After immersing myself in the topic for several years, and speaking to kind and brilliant experts from all over the world, I've realised something. Despite the astonishing complexity of the science of surfaces, we have somehow learned to navigate, and in many cases, control it. Yes, there are still unsolved mysteries and gaps in our knowledge. Yes, there are models that could be improved. And yes, there are disagreements on how even everyday things work (looking at you, curling and adhesives) and on some of the fundamentals. But to focus solely on what we *don't* know would be doing a disservice to what we *do* know.

Our understanding of interfaces has grown alongside us. Throughout history, the more we learned, the wider our horizons became. As a result, very few knowledge gaps have ever truly held us back. We humans are amazingly creative at finding solutions that work, even if we don't have all the equations to hand. The applied knowledge of contact, friction, fluid dynamics and surfaces has allowed us to build pyramids, harness wind energy and explore the solar system. And with each breakthrough we discovered something new, while refining the ideas that had gone before. This is how science and engineering work. It's how they've always worked. And whatever else may come, those pursuits show no sign of stopping.

Further Reading

Ah, references. Simultaneously my favourite and most dreaded part of book-writing. *Sticky* involved a lot of research. My final tally of references – that's the papers, patents, books, articles and reports that I read – is close to 900, and that excludes my interview notes. I won't be sharing all of those here. Instead, what follows is a brief bibliography; a selection of just some of the key references, as well as suggestions for further reading and viewing. You've probably noticed that there are a few titbits in the footnotes, too. However, if you're looking for a specific paper that I've mentioned in the book, and it's not listed below, please check my website. I've published a full reference list there (with links to the papers, where possible): www.lauriewinkless.com

Any other questions? Please contact me on Twitter (@laurie_winkless) and I'll do my best to help.

Introduction: Hello

- Tribology is described as the science of "rubbing and scrubbing": Hähner, G. and Spencer, N. 1998. Rubbing and Scrubbing. *Physics Today* 51, 9: 22.
- Transport of the Colossus: This reproduction of the original mural was made in collaboration with Deir Al-Barsha Youth Union. It was shared on Wikipedia by user Youssef Grace (CC-BY-SA-4.0).

- Fall, A. 2014. Sliding Friction on Wet and Dry Sand. *Physical Review Letters* 112: 175,502.
- Tables of m values can be found in countless textbooks and on lots of websites. These coefficients of static friction were taken from https://www.engineeringtoolbox.com/friction-coefficients-d_778.html

Chapter 1: To Stick or Not to Stick

- Ngarjno, Ungudman, Banggal, Nyawarra and Doring. 2000. *Gwion Gwion: Dulwan Mamaa.* ISBN-10: 38290406: 78.
- An archaeological study on tools found at Minjiwarra – a large sedimentary feature in north-eastern Kimberley – concluded that this region has been continuously occupied by Aboriginal communities for 50,000 years. Veth, P. et al. 2019. Minjiwarra: archaeological evidence of human occupation of Australia's northern Kimberley by 50,000 BP. *Australian Archaeology.* DOI: 10.1080/03122417.2019.1650479
- Finch, D. et al. 2020. 12,000-Year-old Aboriginal rock art from the Kimberley region, Western Australia. *Science Advances* 6: eaay3922. This is also the source of the description of Gwion-style art.
- Coles, D. 2019. *Chromatopia: An Illustrated History of Color.* ISBN-10: 1760760021.
- Kendall, K. 2001. *Molecular Adhesion and its Applications: The Sticky Universe* (151).
- Adhesion models according to 3M™: https://www.3m.com/3M/en_US/bonding-and-assembly-us/resources/how-does-adhesion-work/

- For those of you keen to get guidance on all things adhesive, Steven Abbott has a fantastic website that includes lots of practical guidance (and online calculators) to guide you: https://www.stevenabbott.co.uk/
- You'll find a full list of Barthlott's publications here: http://www.lotus-salvinia.de/index.php/en/publication Most of the early papers are written in German.
- Barthlott, W. and Neinhuis, C. 1997. Purity of the sacred lotus, or escape from contamination in biological surfaces. *Planta* 202: 1–8. This is also the source of Figure 3.

Chapter 2: A Gecko's Grip

- Autumn, K. et al. June 2000. Adhesive force of a single gecko foot-hair. *Nature* 405.
- Ruibal, R. and Ernst, V. 1965. The structure of the digital setae of lizards. *Journal of Morphology* 117, 3.
- Cutkosky, M.R. Climbing with adhesion: from bioinspiration to biounderstanding. *Interface Focus* 5: 20150015.
- The speed of a gecko's foot is taken from Autumn, K. et al. 2006. Dynamics of geckos running vertically. *Journal of Experimental Biology* 209: 260–272. Blinking, according to Harvard, takes between 0.1 to 0.4 seconds, or 100 to 400 milliseconds.
- Autumn, K. et al. 2002. Evidence for van der Waals adhesion in gecko setae. *PNAS* 99, 19: 12,252–12,256. The publication of this paper

led to one of my favourite ever news story headlines: 'How Geckos Stick on der Waals' (*Science*, 2002).

- Stark, A.Y. et al. 2012. The effect of surface water and wetting on gecko adhesion. *Journal of Experimental Biology* 215: 3080–3086.
- The contact angle of water on a gecko toe pad is approximately 150°. Badge, I. et al. 2014. The Role of Surface Chemistry in Adhesion and Wetting of Gecko Toe Pads. *Scientific Reports* 4, article no. 6643.
- Stark, A.Y. et al. 2015. Run don't walk: locomotor performance of geckos on wet substrates. *Journal of Experimental Biology* 218: 2435–2441.
- Much of Hiller's early work was published in German, but this book chapter provides a useful overview: Hiller U.N. 2009. Water Repellence in Gecko Skin: How Do Geckos Keep Clean? in Gorb S.N. (ed.). 2009. *Functional Surfaces in Biology.* Springer, Dordrecht.
- Geim, A.K. et al. July 2003. Microfabricated adhesive mimicking gecko foot-hair. *Nature Materials* 2.
- The first geckos are believed to be at least as old as the dinosaurs. Conrad, J.L., Norell, M.A. 2006. High-resolution X-ray computed tomography of an early Cretaceous gekkonomorph (Squamata) from Öösh (Övörkhangai; Mongolia). *Historical Biology* 18: 405–31.
- Han A.K, et al. 2021. Hybrid electrostatic and gecko-inspired gripping pads for manipulating bulky, non-smooth items. *Smart Materials and*

Structures 30 025010 (9pp). A patent application (US20140272272A1) for a similar, but non-identical, technology was submitted by the Illinois Institute of Technology in 2014. At the time of writing, that application had been abandoned.

Chapter 3: Gone Swimming

- Fairhurst's patents include US6446264 B2 Articles of Clothing, and USD456110 Garment.
- Toussaint, H.M. et al. 2002. Effect of a Fast-skin™ 'Body' Suit on Drag during Front Crawl Swimming. *Swimming, Sports Biomechanics* 1, 1: 1–10.
- Stager, J.M. et al. May 2001. Predicting elite swim performance at the USA 2000 Olympic swim trials. *Medicine & Science in Sports & Exercise* 33, 5:S159; Sanders, R. et al. 2001. Bodysuit yourself: but first think about it. *Journal of Turbulence (Electronic Journal).*
- Oeffner, J. and Lauder, G.V. 2012. The hydrodynamic function of shark skin and two biomimetic applications. *Journal of Experimental Biology* 215: 785–795.
- Some estimates suggest that the drag experienced by the human body moving in water is 780 times larger than the drag experienced when moving in air.
- Statistics on the LZR suit: https://swimswam.com/speedo-fastskin-a-history-of-the-worlds-fastest-swimsuits/

- FINA Requirements for Swimwear Approval https://www.fina.org/sites/default/files/frsa.pdf
- Patent, Holst B. and Akhtar, N. WO 2018/197858 A1. Microstructured sapphire substrates.
- Barthlott, W. et al. 2010. The Salvinia Paradox: Superhydrophobic Surfaces with Hydrophilic Pins for Air Retention under Water. *Advanced Materials* 22: 2,325–2,328.

Chapter 4: Flying High

- Pugh, L.G.C.E. May 1970. Oxygen intake in track and treadmill running with observations on the effect of air resistance. *Journal of Physiology* 207, 3: 823–835; Pugh, L.G.C.E. March 1971. The influence of wind resistance in running and walking and the mechanical efficiency of work against horizontal or vertical forces. *Journal of Physiology* 213, 2: 255–276.
- These values were taken from two fairly niche research papers: 'Effect of moisture content on the viscosity of honey at different temperatures' (*Journal of Food Engineering*, 01 Feb. 2006, 72, 4: 372–377) and 'The rheological properties of ketchup as a function of different hydrocolloids and temperature' (*International Journal of Food Science & Technology*, 44, 3). That's how much I love you.
- Cross, R. 2012. Aerodynamics in the classroom and at the ball park. *American Journal of Physics* 80, 289. Cross also wrote an excellent blog post on the topic: Search for "cross knuckleballs" to find it.

- For a more in-depth history of supersonic flight, plus references to the many studies I've mentioned in this section, I highly recommend reading Chapter 3 of the superbly researched NASA History book *From Engineering Science to Big Science*. The entire book is freely available from here: https://history.nasa.gov/
- NASA Report, *Transiting from Air to Space: The North American X-15*. This is also freely-available on the NASA History website.
- NASA book, *Engineering Innovations*, published online by the Johnson Space Centre. Chapter title: 'Thermal Protection Systems'. Entire book available from (you guessed it) the NASA History site.

Chapter 5: Hit the Road

- Back in 2001, Michelin produced a truly brilliant resource on the fundamentals of tyre grip – Gemma Hatton kindly introduced me to it during my research stage. It's a must-read for anyone who wants to delve even deeper into the topic than I have here. Search for "michelin tire grip dimnp" on any browser to find it.
- I'm lucky enough to live with an expert on road noise – my husband, Richard Jackett, who has carried out lots of research for Waka Kotahi, the New Zealand Transport Agency. Some of his reports can be found here: https://www.researchgate.net/profile/Richard_Jackett/research

- More information on the early *Benz Patent-Motorwagen* project can be found at https://media.daimler.com
- Sugözü, I. et al. 2015. Friction and wear behaviour of ulexite and cashew in automotive brake pads. *Materials and Technology* 49, 5: 751–758; Ganguly, A. and George, R. 2008. Asbestos free friction composition for brake linings. *Bulletin of Materials Science* 31, 1: 19–22.
- Alnaqi, A.A. et al. 2016. Material characterisation of lightweight disc brake rotors. *Proceedings of the Institute of Mechanical Engineers Part L: Journal of Materials: Design and Applications* 232, 7: 555-565.
- Brake manufacturer Brembo has lots more details about F1 brakes on their website.

Chapter 6: These Shaky Isles

- Brace, W.F. and Byerlee, J.D. 1966. Stick-Slip as a Mechanism for Earthquakes. *Science, New Series* 153, 3,739: 990–992.
- Elevated pore-fluid pressure and landslides: Bogaard, T.A. and Greco, R. 2015. Landslide hydrology: from hydrology to pore pressure, *WIREs Water* 3, 3: 439–459; Carey, J.M. et al. 2019. Displacement mechanisms of slow-moving landslides in response to changes in porewater pressure and dynamic stress. *Earth Surface Dynamics* 7: 707–722.
- Wallace, L.M. 2020. Slow Slip Events in New Zealand. *Annual Review of Earth and Planetary*

Sciences 48: 8.1–8.29. The rate quoted was taken from Figure 2.

- Dragert, H. et al. 2001. A Silent Slip Event on the Deeper Cascadia Subduction Interface. *Science* 292: 1521–1528.
- Rogers, G. and Dragert, H. 2003. Episodic Tremor and Slip on the Cascadia Subduction Zone: The Chatter of Silent Slip. *Science* 300: 1942–1943; Obara, K. 2002. Nonvolcanic Deep Tremor Associated with Subduction in Southwest Japan. *Science* 296: 1679–1681.
- Warren-Smith, E. et al. 2019. Episodic stress and fluid pressure cycling in subducting oceanic crust during slow slip. *Nature Geoscience* 12: 475–481.
- Langridge, R. M. et al. 2018. Coseismic Rupture and Preliminary Slip Estimates for the Papatea Fault and its Role in the 2016 Mw 7.8 Kaikōura, New Zealand, Earthquake. *Bulletin of the Seismological Society of America* 108, 3B: 1596–1622.

Chapter 7: Break the Ice

- Rosenberg, B. 2005. Why Is Ice Slippery. *Physics Today* 58, 12: 50.
- Bowden, F.P. and Hughes, T.P. 1939. The mechanism of sliding on ice and snow. *Proceedings of the Royal Society of London A* 172: 280–298.
- Gurney, C. 1949. Surface Forces in Liquids and Solids. *Proceedings of the Physical Society A* 62: 639.

- Weber, B. et al. 2018. Molecular Insight into the Slipperiness of Ice. *Journal of Physical Chemistry Letters* 9, 11: 2838–2842.
- Burridge, H.C. and Linden, P.F. 2016. Questioning the Mpemba effect. *Scientific Reports* 6, article no. 37665.
- Nyberg, H. et al. 2013. The asymmetrical friction mechanism that puts the curl in the curling stone. *Wear* 301, 1–2: 583–589.
- Shegelski, M.R.A. and Lozowski, E. 2016. Pivot-slide model of the motion of a curling rock. *Canadian Journal of Physics* 94: 1305–1309.
- Shegelski, M.R.A. and Lozowski, E. 2018. First principles pivot-slide model of the motion of a curling rock: Qualitative and quantitative predictions. *Cold Regions Science and Technology* 146: 182–186.
- Honkanen, V. et al. 2018. A surface topography analysis of the curling stone curl mechanism. *Scientific Reports* 8: 8123. By the time I read it, the paper was almost a year old, but had somehow slipped through the complex net of alerts that I'd set up for curling-related publications.
- Penner, A.R. 2019. A Scratch-Guide Model for the Motion of a Curling Rock. *Tribology Letters* 67: 35.1–35.13.

Chapter 8: The Human Touch

- For more on how many senses humans have, see this blog post from Dr Steve Draper, a psychologist from the University of Glasgow: https://www.psy.gla.ac.uk/~steve/best/senses.html

- Skedung, L. et al. 2013. Feeling Small: Exploring the Tactile Perception Limits, *Scientific Reports* 3: 2617.
- *The Fingerprint Sourcebook*, produced by the US Department of Justice in 2011.
- Thompson, W. et al. 2017. Forensic Science Assessments: A Quality and Gap Analysis (Latent Fingerprint Examination). Published online by AAAS.
- Champod, C. 2015. Fingerprint identification: advances since the 2009 National Research Council report. *Philosophical Transactions of the Royal Society B* 370: 20140259.
- Liu, X. et al. Measuring contact area in a sliding human finger-pad contact, *Skin Research Technology*: 1–14.
- Tomlinson, S.E. et al. 2013. Human finger friction in contacts with ridged surfaces. *Wear* 301: 330–337.
- Persson, B.N.J. et al. 2013. Contact Mechanics and Friction on Dry and Wet Human Skin. *Tribology Letters* 50: 17–30.
- Changizi, M. et al. 2011. Are Wet-Induced Wrinkled Fingers Primate Rain Treads? *Brain, Behaviour and Evolution* 77, 4: 286–290.
- O'Rian, S. 1973. New and Simple Test of Nerve Function in Hand. *British Medical Journal* 3: 615–616.
- Lederman, S.J. and Klatzky, R.L. 1987. Hand movements: A window into haptic object recognition. *Cognitive Psychology* 19: 342–368.

- Runyan, N.H. and Blaize, D.B. August 2011. The Continuing Quest for the 'Holy Braille' of Tactile Displays. *Proceedings of SPIE 8107, Nano-Opto-Mechanical Systems (NOMS)*: 81070G.
- Russomanno, A. et al. 2015. Refreshing Refreshable Braille Displays. *IEEE Transactions on Haptics* 8, 3: 287–297.
- O'Modhrain, S. July–Sept. 2015. Designing Media for Visually-Impaired Users of Refreshable Touch Displays: Possibilities and Pitfalls. *IEEE Transactions on Haptics* 8, 3: 248–257.
- Culbertson, H. and Kuchenbecker, K.J. 2017. Importance of Matching Physical Friction, Hardness, and Texture in Creating Realistic Haptic Virtual Surfaces. *IEEE Transactions on Haptics* 10, 1: 63–74.

Chapter 9: Close Contact

- *The History of Gauge Blocks*, a short book by precision instrument-makers Mitutoyo was published online in 2015 and is still freely available on the company's website.
- Doiron, T. and Beers, J. 1995. *The Gauge Block Handbook*. Monograph #180 from the National Institute of Standards and Technology (NIST).
- Melzer, M. 2007. *Mission to Jupiter: A History of the Galileo Project*. NASA (SP-2007-4231). Freely available from the NASA History website.
- The Feynman Lectures, 12-2 Friction: '*The reason for this unexpected behavior is that when the atoms in contact are all of the same kind, there is no way for the*

atoms to "know" that they are in different pieces of copper. When there are other atoms, in the oxides and greases and more complicated thin surface layers of contaminants in between, the atoms "know" when they are not on the same part.'

- Lu, Y. et al. 2010. Cold welding of ultrathin gold nanowires. *Nature Nanotechnology* 5: 218–224. This paper is also the source of videos mentioned in the chapter.
- Vazirisereshk, M.R. et al. 2019. Solid Lubrication with MoS_2: A Review. *Lubricants*, 7: 57.
- This paper is a great summary of gold's changing properties: Haruta, M. 2003. When gold is not noble: catalysis by nanoparticles. *The Chemical Record* 3, 2: 75–87.
- Vazirisereshk, M.R. et al. 2019. Origin of Nanoscale Friction Contrast between Supported Graphene, MoS_2, and a Graphene/MoS_2 Heterostructure. *Nano Letters* 19, 8: 5496–5505.
- Dienwiebel, M. et al. 2004. Superlubricity of Graphite. *Physical Review Letters* 92, 12: 126101. The effect had been predicted in the 1980s and the first computational/theoretical study into it was carried out in the early 1990s.
- Erdemir, A. and Martin, J.M. 2018. Superlubricity: Friction's vanishing act. *Physics Today* 71, 4: 40.
- Chen, Z. et al. 2020. Identifying physical and chemical contributions to friction: A comparative study of chemically inert and active graphene step edges. *ACS Applied Materials and Interfaces* 12, 26: 30007–30015.

- Streator, J.L. 2019. Nanoscale Friction: Phonon Contributions for Single and Multiple Contacts. *Frontiers in Mechanical Engineering* 5: 23.
- Stachowiak, G.W. 2017. How tribology has been helping us to advance and to survive. *Friction* 5, 3: 233–247.

Acknowledgements

This book might only have one name on the front cover, but without the wonderful humans listed below, it would never have happened. If you gave up your time to help me in any way, I owe you a debt.

Those I interviewed and quoted: Monique Parsler, Colin Gooch, Gabriel Nodea, Marcelle Scott, Steven Abbott, Adrian Lutey, Kellar Autumn, Alyssa Stark, Mark Cutkosky, Arul Suresh, Aaron Parness, Amy Kyungwon Han, Fiona Fairhurst, Melissa Cristina Márquez, Dylan Wainwright, Bodil Holst, Maz Jovanovich, Andrew Neely, Priyanka Dhopade, Jon Marshall, Gemma Hatton, Shahriar Kosarieh, John Carey, Carolyn Boulton, Laura Wallace, Jeremy Gosselin, Rob Langridge, Emily Warren-Smith, Daniel Bonn, Mark Shegelski, Staffan Jacobson, Mark Callan, Christina Hulbe, Amy Betz, Gilane Khalil, Tanja Van Peer, Matt Carré, Sile O'Modhrain, Heather Culbertson, Lenice Evergreen, Nina Wronski, Ashlie Martini, Philip Moriarty, Jeffrey L. Streator.

Those who advised me, talked about their work, introduced me to others, lined up interviews, sent me reports, papers, images and data, or fact-checked my ramblings: Geoff Wilmott, Chiara Neto, Emile Webster, Jenny Malmström, the team at NZ Police (Matt, Eugene, Tony and Greg), James Batteas, Dan Bernasconi, Wilhelm Bathlott, Alex Russomanno, Rachel Oliver, Alan Baxter, Hannah Davidson, Geoff Kilgour, Robyn Sloggett, Zhū Creative, Will Hings, Bettina Mears, Maggie McArthur Murray, Vanessa Young, Jim Sutton Charles Dhong, Mark Lincoln and Adam Parr.

My chapter reviewers: Lisa Martin, John Uhlrich, Eleanor Schofield, Laura Sessions, Paul Byrne, Clare Bardsley, Leanne Ross, David Lamb, Nic Harrigan, Celeste Skatchill, Catherine Qualtrough, Rebecca O'Hare, Anna Sampson, Gareth Hinds, Felicity Powell, @ThatMaoriGirl, Nicola Hardy, Claire Lucas, Orla Wilson, John Englishby.

My husband, whose patience is unmatched, and who has lived through every single word of this book. My parents and family who never doubted that I'd eventually finish writing it, and who tell me that they're proud of me, even when I'm a mess. My friends, scattered across multiple timezones, who are my main source of moral support and welcome distraction, especially when things are wobbly. My Sigma-siblings for continuously lifting me up. My Twitter-fam and occasional friendly stranger for basically being my colleagues, cheerleaders and confidants. And Mary Roach for giving me the thumbs-up to use a book title that she inspired.

... And finally, a big thanks to the team at **Bloomsbury**, especially Sigma-Daddy Jim Martin, who didn't shout at me when I moved to the other side of the world, nor when I told him the book was going to be >1.5 years late. Catherine Best and Marc Dando for helping to turn a rough draft into something worth publishing. Anna MacDiarmid who set *Sticky* on its long path, and Angelique Neumann who carried it home.

Thank you all x

Index

Abbott, Steven 32, 33, 37–38, 39, 48
ablative materials 146–147
Aboriginal rock art 17–24
acrylate spheres 40–41
acrylic polymers 28–29
adhesion 30–34
  chemical 31
  crack energy and 37–38
  diffusion 32–33, 288
  electrostatic interactions 33–34, 55–56, 64, 81–82
  mechanical 31–32
  molecular 158–159, 164
  surface energy and 34–38
  van der Waals forces 33–34, 62–70, 74, 81
  see also gecko adhesion
adhesives 30, 31, 33, 34, 36–43
  gecko-inspired 71, 75–78, 81–82
  Post-It Notes 39–41
  superglue 41–43
  surface energy and 36–38
Adidas 85, 88, 92, 95
aerodynamics 117–147
  balls 123–132
  downforce in Formula 1 cars 151–152
  drag 117–123, 126–127, 139–140, 142–147
  frictional heating 142–147
  hypersonic flight 144–147
  laminar and turbulent flow 91, 120–123, 126–128, 130, 132, 138
  sound barrier 135–140
  speed of sound 134–135
  supersonic flight 133–144
  transonic instability 138–142
air lubrication systems 111–114
aircraft
  frictional heating 142–145
  hypersonic flight 144–145
  ice formation 244–246
  reduced-gravity 79, 80
  rocket-powered 144–145
  supersonic flight 133–144
  transonic instability 138–142
Aizenberg, Joanna 109–110
aluminium brakes 175–176
Amontons, Guillaume 308
amphiphilic molecules 29
anti-fouling paints 109–111
anti-microbial surfaces 46–47
Apollo missions 145, 146
Aristotle 308
asbestos 169
Assisted Tow Method (ATM) 94
asymmetric friction 228
atomic force microscopy (AFM) 292–293, 294–297, 298–300, 301–303, 305
Autumn, Kellar 53, 54, 56, 59–60, 65, 70, 73

balls 123–132
Barthlott, Wilhelm 43–44, 112–113, 114
baseball 129, 131–132
Batteas, James 303
Beckham, David 128–129
Bell X-1 (aircraft) 133–134, 135, 140–142
Benjanuvatra, Nat 96
Benz, Bertha 166–169
Benz, Carl 167, 168
Benz Patent-Motorwagen 166–169
Bernasconi, Dan 113–114

Betz, Amy 242–245
binders
  brake pads 169, 171–172
  paint 19, 26, 28–29, 31
  Post-It Notes 40–41
biocides 109
biofilms 46–47
biofouling 107, 109–111, 113
biphilic surfaces 243–245
boats and ships 106–115
  air lubrication systems 111–114
  biofouling 107, 109–111, 113
Boehler, Remy 223
Bonn, Daniel 12–14, 216–220
Bonn, Mischa 216–220
Boulton, Carolyn 197–198, 199–200
boundary layers 120–123, 126–128, 130, 132, 138, 142–143
bow shock 141, 146
Brace, William F. 186–187
braille 265, 268–274
brake fade 171
brakes 168–177
Britten, Jack 20–21
buoyancy 90, 105–106
Burns, Robert 225
Byerlee, James D. 186–187

Callan, Mark 224
Cappaert, Jane 103–104
carbon-carbon 176
Carey, Jon 188, 191–197
Carpick, Robert 295, 306
Carré, Matt 256, 257, 258–259
Cascadia subduction zone 202, 204, 206
Champod, Christophe 255
Chan, Wai Pang 60
chemical adhesion 31
Christchurch earthquakes 192–193
climate change 114–115, 240
coefficient of friction 12–13, 36, 48, 164, 174, 176, 198, 217, 257, 260, 301, 309–311
cohesion 34
cold welding 285–289
Coles, David 25–26
Concorde (aircraft) 142, 143
Coover, Harry 42
crack energy 37–38
cricket balls 129, 130–131
Crimmen, Oliver 86–87
Cross, Rod 132
Culbertson, Heather 275–278
curing reactions
  brake pads 171
  paint 26
  superglue 42
  tyre rubber 156
curling 225–237
Cutkosky, Mark 72–75, 76, 77–78
cyanoacrylate adhesives 41–43

de Bruijn, Inge 86
degressive springs 76
de-icing fluids 244
Dellit, Wolf-Dietrich 53, 55
dermal denticles 86–87, 99, 100–101, 102–103
desert varnish 23–24
Dhopade, Priyanka 143, 146
diffraction limit 57
diffusion 32–33, 288
Digital Particle Image Velocimetry (DPIV) 102
dipoles 62–64
displacement hulls 107
dissociation of air 146
Djhutyhotep mural, Egypt 11–14
dolphins 86, 101
downforce 151–152
drag forces
  aerodynamic drag 117–123, 126–127, 139–140, 142–147
  form drag 89–90, 92, 104, 107, 108, 119, 126–127
  frictional heating 142–147
  pressure drag 119, 126–127
  rolling drag 162–163
  skin friction 91–92, 107, 108–109, 113, 119, 123, 126, 127, 142–147

wave drag 90–91, 104, 107, 108, 139–140, 142
see also hydrodynamic drag
Dragert, Herb 202, 204
Dudhia, Anu 108
dynamic viscosity 117–118, 122

earthquakes 179, 181, 182–211
depth 189–190
episodic tremors 204
fault behaviour during 197–200
fault types 182–184, 190
Kaikōura earthquake 207–210
lab-based experiments 190–197
liquefaction 192–193
moment magnitude 188–190
pore-fluid pressure 191–193, 195, 205–206
seismic waves 184, 188, 189–190, 204
slow-slip events (SSEs) 202–207, 208
stick-slip motion 184–188
subduction zones 179, 190, 200–207
elasticity 153–154
elastomer actuators 78
electrons 55, 57–58, 63–64, 302, 303–305
electrostatic chucks 81
electrostatic interactions 33–34, 55–56, 64, 81–82
episodic tremors 204
Ernst, Valerie 58, 59
Escherichia coli (*E. coli*) 46–47
Evergreen, Lenice 279–280, 281, 282–283

Fairhurst, Fiona 86–88, 89, 99, 103–104
Faraday, Michael 215
FASTSKIN suit 87–88, 95–96, 99, 101–103, 104
Faulds, Henry 252
faults 182–184
behaviour during earthquakes 197–200
episodic tremors 204
friction 184–188, 196, 198–200, 201
gouge materials 198–200
lab-based experiments 190–197
normal faults 183
Papatea Fault 208–210
pore-fluid pressure 191–193, 195, 205–206
shear forces 195–196
slow-slip events (SSEs) 202–207, 208
stick-slip motion 184–188
strike-slip faults 183, 184, 190
thrust faults 183, 190
Feynman, Richard 286
FINA (International Swimming Federation) 98, 106
fingerprints 249–256
fluid dynamics 14, 119–123
forensic science 253–256
form drag 89–90, 92, 104, 107, 108, 119, 126–127
Formula 1 cars 149–152
brakes 170, 172–174, 176–177
downforce 151–152
tyres 153, 155–157, 159–162, 163–164, 165
friction 11–14, 308–309
asymmetric 228
atomic 292–301, 304–307, 309–311
brakes 168–177
chemical 301
coefficient of 12–13, 36, 48, 164, 174, 176, 198, 217, 257, 260, 301, 309–311
faults 184–188, 196, 198–200, 201
human skin and 256–261
lubrication and 11–14, 290–294
phonons 305–307, 310–311
physical 301
rate and state laws 198–200
rolling resistance 162–163
sliding/kinetic 11–14, 216–220, 305, 310

static 11, 54, 185, 198
stick-slip 184–188, 295–296
structural superlubricity 297–300
tyre grip mechanisms 157–159, 164, 165
see also drag forces
friction modifiers 169
friction ridges *see* papillary ridges
frictional heating
aircraft 142–145
brakes 170–172, 173–174, 175
phonons 305–307, 310–311
re-entry vehicles 145–147
frictional melting 214–215
Fry, Arthur 40–41

Galileo spacecraft 284–286
Galton, Francis 252–253
gauge blocks 279–283, 284
gecko adhesion 51–83
adhesives, gecko-inspired 71, 75–78, 81–82
directionality of 61, 64–65
electrostatic interactions 55–56, 64, 81–82
lamellae 52–54, 58–59, 64, 67, 73
micro-interlocking 54
robotics, gecko-inspired 72–75, 78–82
setae 54, 56, 58–59, 60, 61–62, 64–66, 67–68, 69–70, 73, 74, 75, 79
spatulae 58–59, 61, 62–63, 64, 67–68, 69, 70, 73, 79
static friction 54
suction 52–53
Teflon and 68–69
van der Waals forces 62–70, 74, 81
water and 66–70
gecko tape 75–78, 81–82
Geim, Andre 71
geological processes 179–182
see also earthquakes; faults; glaciers; tectonic plates
GeoNet 181, 189, 201–202, 203
Gija people 19–22
Gillespie, Brent 271–272
glaciers 237–240
golf balls 124–127
Gooch, Colin 26–30
Goodwin, Dan 52
Gosselin, Jeremy 204, 206
gouge materials 198–200
graphene 293–294, 295, 298–300
graphite 291–292, 293, 295, 296–297, 298–300
greenhouse gas emissions 114–115
gripper devices 78–82
Gurney, Charles 215–216
Gwion artwork 18–19, 21

Han, Amy Kyungwon 81–82
haptography 275–278
Hatton, Gemma 153, 155, 157, 159–161
Hawkes, Elliot 75–76, 77, 80
Hawkins, Simon 18
Helfand, Eugene 33
Henry, Edward 252, 253
Hikurangi subduction zone 190, 200, 204–205
Hiller, Uwe 68
HOBITSS experiment 205, 206
Holst, Bodil 110–111
Hulbe, Christina 238–239, 240
hydrodynamic drag 85–115
boats and ships 106–115
measurement 92, 93–97
sharks 86–87, 99–103
swimsuits 85–88, 95–106
types 89–92
hydrophilic attraction 28, 29, 35, 71, 113, 243
hydrophobic repulsion 29, 35–36, 44, 45, 47, 67–69, 113, 243
hypersonic flight 144–147
hysteresis 154, 157, 259

ice 213–246
curling 225–237
formation 240–246

frictional melting 214–215
glaciers 237–240
nucleation 241–242
pressure melting 213–215
skating 213–214, 220, 221–224
sliding friction 216–220
surface molecules 215–220
indentation, of tyres 157, 159, 164, 165
interlocking 31–32
International Space Station 80, 81

Jacobson, Staffan 230–231, 232, 236
Johansson, Carl Edvard 280, 281–282
Johnson, Charlie 295
Jovanovich, Marija 136, 138–139, 142
Kaikōura earthquake 207–210
Katz, David 269–270
Kendall, Kevin 31–32, 38
keratins 59, 69–70, 250
Khalil, Gilane 249–250
Kim, Seong H. 298–300
kinetic friction 11–14, 216–220, 305, 310
Klim, Michael 86
Knight, Pete 144–145
knuckleball pitching 131–132
Koala 252
Kosarieh, Shahriar 174, 175–176

laminar and turbulent flow 91, 120–123, 126–128, 130, 132, 138
Langridge, Rob 208–209
laser-induced periodic surface structures (LIPSS) 46–47
Lauder, George 99, 101–103
Leonardo da Vinci 308
Linden, David J. 247, 250
linseed oil 25, 26
lipids 70
liquefaction 192–193
lotus effect 43–46
Lou, Jun 287–289
Lozowski, Edward 233–235, 237
lubrication 11–14, 290–294
air lubrication systems 111–114
structural superlubricity 297–300
Lutey, Adrian 46–47
Lyndsell, Barbara 194–195
LZR Racer suit 105–106

MAD (Measuring Active Drag) system 93–94, 95–96
Maderson, Paul 59
Maersk 114–115
Magnus effect 128–129
Márquez, Melissa Cristina 100
Marshall, Jonathan 149–150, 173, 174, 176–177
Martini, Ashlie 289, 290, 291–292, 294–295, 297, 298–300, 305, 311
mechanical adhesion 31–32
mechanoreceptors 263–266, 269, 270, 274–275, 276
Meissner corpuscles 264–265, 269
Merkel discs 263–264, 269
Miskew, Emma 231
molecular adhesion 158–159, 164
molybdenum disulfide 291–292, 293, 294, 295, 296
moment magnitude 188–190
Moriarty, Philip 302–303, 304

Nathan, Alan M. 132
Neely, Andrew 143, 144, 146
Nepenthes 109
Nodea, Gabriel 19–20, 21–22
non-volcanic tremors 204
Novoselov, Konstantin 71
nucleation 241–242

ochre 17–24
Olympic Games 85–86, 87–88, 97, 98–99, 105–106, 221, 223, 225
O'Modhrain, Sile 267–268, 269–274
opinion evidence 254–255

Pacinian corpuscles 265, 274–275, 276
paint 24–30, 31, 34, 36
  anti-fouling 109–111
  see also ochre
Papatea Fault 208–210
papillary ridges 249–266
  fingerprints 249–256
  frictional interactions and 256–261
  mechanoreceptors and 263–266, 269
  water-induced wrinkles and 261–262
Park, Roseleen 21
Parness, Aaron 79–81
Parsler, Monique 34, 36–37
Pauli exclusion principle 302–303
Penner, A. Raymond 236–237
Phelps, Michael 88
phonons 305–307, 310–311
pigments 24, 25–26, 27–29, 31
  see also ochre
pitcher plants 109
plasma electrolytic oxidation (PEO) 175–176
plate tectonics *see* tectonic plates
polymers
  acrylate spheres 40–41
  acrylics 28–29
  cyanoacrylates 41–43
  diffusion 32–33
  silicone-based 82, 110
  Teflon 47–50, 68–69, 85
  tyre rubber 153–154, 156
polyurethane swimsuits 105–106
pore-fluid pressure 191–193, 195, 205–206
Post-It Notes 39–41
pressure drag 119, 126–127
pressure melting 213–215
primates 251–252
Pugh, Griffith 118

Ratvasky, Thomas 245
Red Hands Cave, Australia 24
re-entry vehicles 145–147
regelation 215
Resene 26–30
resins 28–29
Reynolds number 121–123
Richter scale 188
Ridley, Jack 133
Right Stuff, The (film) 133, 135, 142
robotic systems, gecko-inspired 72–75, 78–83
rock art 17–24
Rogers, Garry 204
rolling resistance 162–163
Rosenberg, Bob 214
rowing shells 107–108
Ruffini endings 265–266
Ruibal, Rodolfo 58, 59
Rushall, Brent 97, 98
Russomanno, Alexander 271–272

Salvinia fern 112–113
saponite 198, 199–200
scanning electron microscopy (SEM) 43–44, 57–58
Schimmel, Thomas 112–113, 114
Schlieren photography 138
Scott, Marcelle 20, 22, 23
self-cleaning surfaces 43–46, 110–111
sharks 86–87, 99–103
Shegelski, Mark 229–230, 231–236, 237
shipping *see* boats and ships
shockwaves 138–139, 141–142, 146
Silver, Spencer 39–41
skating 213–214, 220, 221–224
skin, human 247–266
  fingerprints 249–256
  frictional interactions 256–261
  glabrous 248–249
  mechanoreceptors 263–266, 269, 270, 274–275, 276
  viscoelasticity 257, 258–259, 261
  water-induced wrinkles 261–262

skin friction 91–92, 107, 108–109, 113, 119, 123, 126, 127, 142–147
sliding friction 11–14, 216–220, 305, 310
SLIPS (slippery liquid-infused porous surfaces) 109–110
slow-slip events (SSEs) 202–207, 208
sound, speed of 134–135
sound barrier 135–140
space programme
  Galileo spacecraft 284–286
  gripper devices 79–81
  re-entry vehicles 145–147
Space Shuttle 145, 146, 176, 285–286
speed of sound 134–135
Speedo 86, 87–88, 92, 95–96, 99, 104, 105–106
Stager, Joel 96–97
Staphylococcus aureus (*S. aureus*) 46–47
Stark, Alyssa 56, 57, 65–69, 70
stick-slip friction 184–188, 295–296
Stickybot 73–75
Streator, Jeffrey L. 306–307, 310
structural superlubricity 297–300
subduction zones 179, 190, 200–207
suction 52–53
superglue 41–43
superhydrophobic repulsion 44, 47, 67–69
supersonic flight 133–144
Suresh, Arul 76–77, 78
surface drag *see* skin friction
surface energy 34–38, 47, 48, 67–69
surface relaxation 288
surface-on-surface contact 279–311
  atomic force microscopy (AFM) 292–293, 294–297, 298–300, 301–303, 305
  atomic friction 292–301, 304–307, 309–311
  cold welding 285–289
  defining 301–304
  diffusion 288
  gauge blocks 279–283, 284
  lubrication 11–14, 290–294
  phonons 305–307, 310–311
  structural superlubricity 297–300
  surface relaxation 288
surfactants 29
sweat 260
sweat glands 250–251, 262
swimming 85–106
  drag force measurement 92, 93–97
  drag force types 89–92
  propulsive forces 88–89
  sharks 86–87, 99–103
  swimsuits 85–88, 95–106
swing bowling 130–131
Symonds, Pat 173–174

Taylor, William 124, 127
tectonic plates 179–184
  convergent boundaries 180–181, 183, 190, 200–207
  divergent boundaries 180, 183
  subduction zones 179, 190, 200–207
  transform boundaries 180–181, 184, 190, 203
Teflon 47–50, 68–69, 85
temperature
  brakes 170–172, 173–174, 175
  friction of ice surfaces and 217–220
  of ice for winter sports 213–214, 223, 224
  ice formation and 240–241, 242, 245
  speed of sound and 134–135
  tyre grip and 159–161
  see also frictional heating
Thorpe, Ian 85, 89
titanium dioxide 25, 27–28

Toet, Willem 151–152
Toft, Soren 114–115
touch 247–249, 256–278
  active 270
  braille 265, 268–274
  frictional interactions 256–261
  haptography 275–278
  mechanoreceptors 263–266, 269, 270, 274–275, 276
  tool-mediated 274–275
  water-induced wrinkles and 261–262
Toussaint, Huub 95–96
transmission electron microscopy (TEM) 287–289
transonic instability 138–142
Transport of the Colossus mural, Egypt 11–14
TresClean 46–47
tribofilms 174–175
tribology 10–11, 256, 290, 305
triple-roll milling 25–26
turbulence *see* laminar and turbulent flow
tyres 152–166
  grip mechanisms 157–159, 164, 165
  materials 153–157
  noise 165–166
  rolling resistance 162–163
  temperature and 159–161
  tread patterns 155–156, 163–166

van der Waals forces 33–34, 62–70, 74, 81, 158, 283, 302–303
van Leeuwen, Hans 214
Van Peer, Tanja 254, 255
Velocity Perturbation Method (VPM) 94
violins 185–186
viscoelasticity 153–154, 157, 158, 162–163, 257, 258–259, 261
viscosity 111, 117–118, 121–122, 154
visual impairment 265, 266, 267–274
volcanic eruption 181–182
vulcanisation 156

Wagler, Johann 52
Wainwright, Dylan 101
Wallace, Laura 190–191, 200–201, 202, 203, 204–205, 206–207, 208
Warmun Art Centre, Australia 21–22
Warren-Smith, Emily 205–206
water
  contact angle of 35–36, 44, 45
  gecko adhesion and 66–70
  human skin and 259–262
  hydrophilic attraction 28, 29, 35, 71, 113, 243
  hydrophobic repulsion 29, 35–36, 44, 45, 47, 67–69, 113, 243
  tyre grip and 158, 164–165
  see also hydrodynamic drag
wave drag 90–91, 104, 107, 108, 139–140, 142
welding, cold 285–289
wettability 34–38, 47, 48, 67–69
Whakaari (White Island), New Zealand 181–182
Wolfe, Tom 133
Wronski, Nina 281, 282, 283

X-15A-2 (aircraft) 144–145

yacht racing 108, 113–114
Yeager, Chuck 133–134, 135–136, 141
Young's modulus 234